Introducing the calculator

Number display

Here is a drawing of a simple calculator.

Switch on your own calculator. You will see 0 (zero) on the number display.

Switch on →	0
Press [8] →	8
Press [C] to clear →	0
Press [7] then [8] →	78
Press [C] to clear →	0

1 Press keys to show the following numbers on the number display. Remember to press the [C] key before you enter the next number.

(a) 12, 34, 56, 78, 91, 80
(b) 123, 456, 789, 321, 654, 987
(c) 1234, 2345, 3456, 4567, 5678, 6789
(d) 105, 2104, 1096, 2040, 3007, 9001

2 Press keys to show the following numbers on the number display. Remember to press the [C] key before you enter the next number.

(a) One hundred and twenty-five
(b) Four hundred and sixty
(c) Two thousand four hundred and fifty
(d) Four thousand and twenty
(e) Ten thousand and sixty-four
(f) Twenty-six thousand and two

3 What is the largest number you can show in the display?

Addition

Find the [+] key and the [=] key.

To find 73 + 62
Press *in order* [7] [3] [+] [6] [2] [=]
The answer 135 is shown on the number display → [135]
Press the clear key [C]
Check your answer by finding 62 + 73.
To do this: press in order [6] [2] [+] [7] [3] [=]

1 Calculate the following. Remember to check your answers.

(a) 37 + 23 (b) 42 + 18 (c) 16 + 44 (d) 25 + 35
(e) 53 + 35 (f) 96 + 44 (g) 69 + 69 (h) 49 + 51
(i) 90 + 70 (j) 36 + 79 (k) 20 + 90 (l) 19 + 91

To find 112 + 124 + 125
Press *in order* [1] [1] [2] [+] [1] [2] [4] [+] [1] [2] [5] [=]
The answer 361 is shown on the number display → [361]
Press the clear key [C]
Check your answer by finding 125 + 124 + 112.

2 Calculate the following. Remember to check your answers.

(a) 73 + 96 + 118 (b) 48 + 39 + 57 (c) 103 + 301 + 130 (d) 907 + 304 + 64

(e) 765 + 49
(f) 386 + 79
(g) 582 + 439
(h) 321 + 456 + 789
(i) 109 + 370 + 640
(j) 1005 + 907 + 1300

3 The table shows the number of spectators at three football matches.

	Men	*Women*	*Children*
Airdrie	2154	176	298
Ayr	5321	197	1423
Falkirk	1204	105	391

Find:
(a) the total number at each game
(b) the total number of **men** at **all** the games
(c) the total number of **spectators** at **all** the games.

4 How far is it from Glasgow to London:
(a) via Hull
(b) via Birmingham?

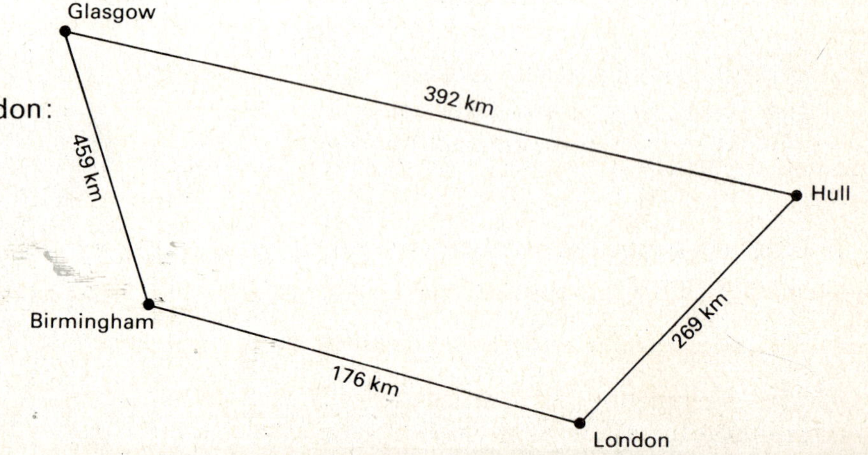

Addition of money – using the [·] key

69p can be written as £0.69
To find 69p + £1.49
Press *in order* [0] [·] [6] [9] [+] [1] [·] [4] [9] [=]
The answer £2.18 is shown on the number display as → [2.18]
Press the clear key [C]
Check your answer by finding £1.49 + 69p.

1 Calculate the following. Remember to check your answers.
(a) 49p + £1.39 (b) £2.18 + 74p (c) £19.37 + 98p (d) £4.04 + £1.06

2 Here is a menu from Louie's cafe.

How much would it cost altogether for:
(a) soup, omelette, trifle, and a glass of milk
(b) juice, cold meat, and coffee
(c) soup, steak, ice-cream, and tea
(d) juice, chicken, trifle, and lemonade?

Write down what you would choose from this menu and find out how much it would cost.

3 Look at the prices of the clothes in the shop window.

Find the cost of:
(a) a T-shirt and a pair of jeans
(b) a football strip and a shirt
(c) all the articles which cost less than £5 each
(d) a shirt and a tie.

4 Look at this poster for a holiday excursion.

How much would it cost for:
(a) one adult and one child
(b) father, mother, one child, and one old age pensioner
(c) one adult, two old age pensioners, and two children?

Subtraction

Find the [–] key.

To find 92 – 76
Press *in order* [9] [2] [–] [7] [6] [=]
The answer 16 is shown on the number display ⟶ [16]

1 Calculate the following.

(a) 60 – 37 (b) 60 – 45 (c) 60 – 12 (d) 60 – 56 (e) 60 – 9
(f) 97 – 39 (g) 146 – 49 (h) 139 – 86 (i) 769 – 89 (j) 364 – 129

2

(a)	(b)	(c)	(d)	(e)
108	365	1000	3456	3015
– 79	– 109	– 846	– 789	– 1090
___	___	___	___	___

You can check your answer to a subtraction by doing an addition as follows.
If you try 362 – 197 the answer in the display should be [165]

To check this answer add these two numbers

$$\begin{array}{r} 362 \\ -\mathbf{197} \\ \hline \mathbf{165} \end{array}$$

You should get [362], the number you started with.

3 Calculate the following. Remember to check your answer by adding.

(a)	(b)	(c)	(d)	(e)
1245	3156	4000	2010	6286
– 679	– 1009	– 3259	– 1509	– 4090
___	___	___	___	___

4 The population of Hamilton is 46 002 and Motherwell 73 564.

Find (a) the total population of the two towns
(b) the difference between the populations.

5 The map shows the populations of Scotland, England, and Wales.

(a) How many more people live in Scotland than in Wales?
(b) Find the difference in population between England and Scotland.
(c) What is the total population of England, Scotland, and Wales?

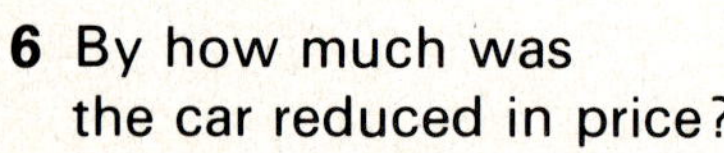

6 By how much was the car reduced in price?

7 Complete the following:

(a) 137 + [ ] = 296 (b) 1937 + [ ] = 3000
(c) [ ] + 349 = 600 (d) [ ] + 9215 = 12 000

Subtraction of money – using the [·] key

To find £5 – £3.75
Press *in order* [5] [–] [3] [·] [7] [5] [=]
The answer £1.25 is shown on the number display as ⟶ [1.25]
Check your answer by adding £1.25 and £3.75

1 Find (a) £9.53 – £6.94 (b) £11.76 – £9.89

2 John saved £21.50 and June saved £17.39. How much more has John saved than June?

3 Helen spends £71.25 on her holiday and Stephen £53.95
How much less did Stephen spend than Helen?

4 A television set is priced at £83.65 in one shop and £89.27 in another.
What is the difference in price?

5 What is your change from £20 if you spend
(a) £11.75 (b) £15.65 (c) £13.37 (d) £9.94?

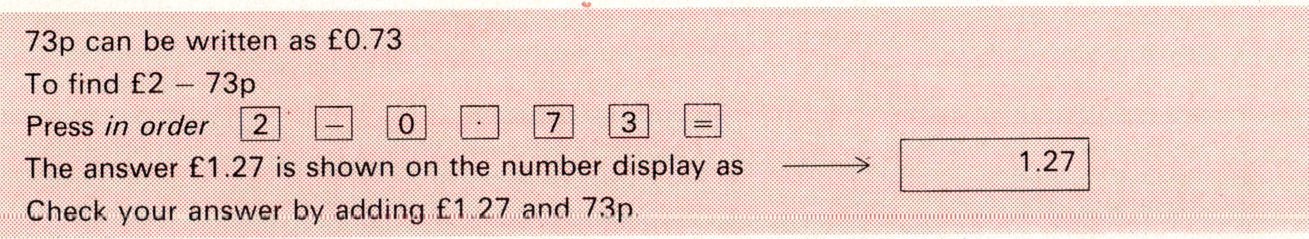

73p can be written as £0.73
To find £2 – 73p
Press *in order* [2] [–] [0] [·] [7] [3] [=]
The answer £1.27 is shown on the number display as ⟶ [1.27]
Check your answer by adding £1.27 and 73p

6 Calculate the following. Remember to check your answer by adding.
(a) £5 – 37p (b) £10 – 49p (c) £3 – 92p (d) £2 – 9p

7 (a) A boy saves £9.32 and his father gives him £7.50. How much has he altogether?
(b) If he now spends £10.40 how much has he left?

8 A boy puts £9.50 into his bank account. Later he puts in £8.25. He then withdraws £15.
How much has he left in his account?

9 A tape recorder costs £15.25 and a girl pays £4.50 deposit.
How much more has she still to pay?

10 Calculate: (a) £11.23 + £9.48 – £16.45
(b) £11.23 – £16.45 + £9.48
What did you notice in your number display when you pressed the [+] key in (b)?

11

(a) Which two articles can you buy if you have £24?
(b) How much money will you have left?

Multiplication

Find the [×] key

To find 12 × 72
Press *in order* [1] [2] [×] [7] [2] [=]
The answer 864 is shown on the number display ⟶ [864]
Press the [C] key.
Check your answer by finding 72 × 12.

1 Calculate the following. Remember to check your answers.
(a) 163 × 25 (b) 47 × 32 (c) 125 × 49 (d) 97 × 97 (e) 1003 × 91

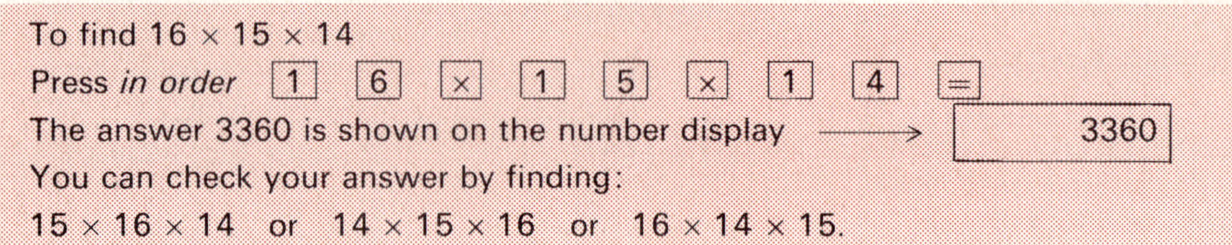

To find 16 × 15 × 14
Press *in order* [1] [6] [×] [1] [5] [×] [1] [4] [=]
The answer 3360 is shown on the number display ⟶ [3360]
You can check your answer by finding:
15 × 16 × 14 or 14 × 15 × 16 or 16 × 14 × 15.

2 Calculate the following. Remember to check your answers.
(a) 17 × 18 × 19 (b) 101 × 102 × 103 (c) 91 × 19 × 17 (d) 35 × 35 × 35

3 Approximately how many matches will be in:
(a) a dozen boxes
(b) a score of boxes
(c) a gross of boxes?

Why are your answers approximate?

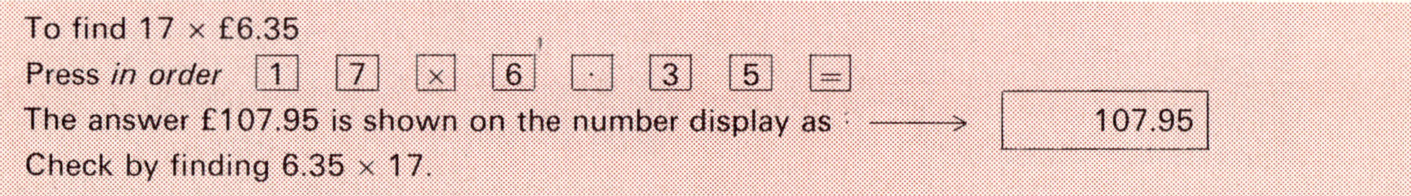

To find 17 × £6.35
Press *in order* [1] [7] [×] [6] [·] [3] [5] [=]
The answer £107.95 is shown on the number display as ⟶ [107.95]
Check by finding 6.35 × 17.

4 Calculate the following. Remember to check your answers.
(a) 23 × £4.65 (b) 19 × £7.28 (c) 37 × £9.38 (d) 40 × £10.55

5 Here is a pay slip showing John's weekly wage. How much does he get in a year?

6 How much does it cost to tile a rectangular hall which is 27 m long and 15 m wide?

Division

Find the [÷] key

To find 351 ÷ 9
Press *in order* [3] [5] [1] [÷] [9] [=]
The answer 39 is shown on the number display ⟶ [39]
Press the [C] key.
Check your answer by doing the division again.

1 Calculate the following. Check your answers.
(a) 392 ÷ 7 (b) 891 ÷ 9 (c) 3648 ÷ 8 (d) 1530 ÷ 10
(e) 817 ÷ 19 (f) 817 ÷ 43 (g) 1508 ÷ 29 (h) 1369 ÷ 37

2 (a) How many 47-seater buses are needed to take 705 children and teachers on an outing?
(b) How many buses will be needed if 16 parents also go on the outing?

3 A group of 27 people win £97 875.
How much do they each get if they all get the same share?

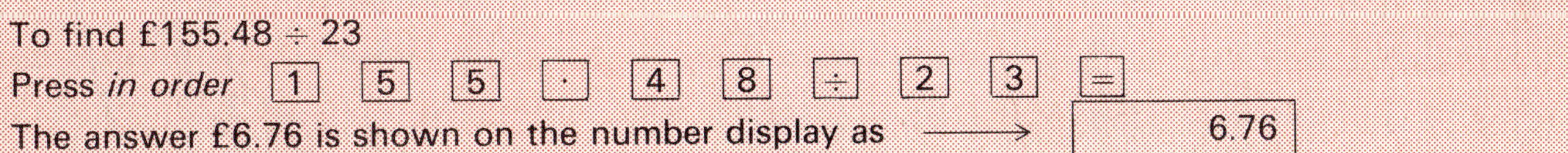
To find £155.48 ÷ 23
Press *in order* [1] [5] [5] [·] [4] [8] [÷] [2] [3] [=]
The answer £6.76 is shown on the number display as ⟶ [6.76]

4 Calculate the following. Remember to check your answer.
(a) £28.98 ÷ 23 (b) £402.96 ÷ 73 (c) £67.83 ÷ 119

5 Anne's yearly pay was £4651.80. How much was this per month?

6 A farmer bought 53 sheep for £5154.25. How much did each sheep cost?

7 A shopkeeper buys 9 boxes of apples costing £5.04 per box.
(a) How much did it cost him altogether?
(b) If he shares the cost equally with his five brothers, how much does each brother owe him?

8 To buy 3 radios for the school, 573 children each pay 18p.
(a) How much is this altogether?
(b) How much does each radio cost?

9 A school collected £262.15 to buy cassette recorders costing £37.45 each.
How many can be bought?

Rounding off

Do not use a calculator for questions 1, 2, 3, and 4.

19 is **nearly** 20 and 12 is **just over** 10.
So 19 × 12 is **approximately** 20 × 10 = 200.
If you use your calculator the answer is 228 – **just over** 200.

1 Copy and complete.

(a) 18 × 22 is about 20 × 20 = 400
(b) 52 × 69 is about __ × __ = ____
(c) 99 × 43 is about __ × __ = ____
(d) 28 × 97 is about __ × __ = ____

2 Write down an approximate answer for each of the following.

(a) 29 × 19 (b) 68 × 41 (c) 96 × 49 (d) 76 × 38

3 Three answers are given to each of the following calculations.
Find an approximate answer.
Then write down the answer you **think** to be correct.

Calculation			
(a) 49 × 57 ⟶	27933	2793	273
(b) 23 × 38 ⟶	84	874	8744
(c) 77 × 33 ⟶	2541	251	25411
(d) 21 × 99 ⟶	20799	207909	2079

678 is **nearer** to 700 than 600.
678 **rounded off** to the nearest hundred is 700.
2342 **rounded off** to the nearest hundred is 2300.

4 Round off the following numbers to the nearest hundred.
(a) 814 (b) 576 (c) 984 (d) 1234 (e) 2359

Using a calculator 63 × 48 = 3024
Rounded off to the nearest hundred the answer is 3000.

5 Trace this diagram of dots and numbers.

6 Using a calculator, find the answers to the following and round off each answer to the nearest hundred.

(a) 63 × 48	(b) 46 × 32	(c) 97 × 21
(d) 65 × 19	(e) 83 × 29	(f) 26 × 63
(g) 109 × 17	(h) 128 × 17	(i) 47 × 53

7 Using a ruler, join the dots between the answers to question 6 (a) and (b), (b) and (c), (c) and (d), and so on.

Fun with numbers

1 Copy and complete $37 \times 3 =$
$37 \times 6 =$
$37 \times 9 =$
$37 \times 12 =$

What happens when you continue to multiply 37 by multiples of 3?
Try up to 27.

2 Copy and complete $999\,999 \times 2 =$
$999\,999 \times 3 =$
$999\,999 \times 4 =$
$\vdots$
$999\,999 \times 9 =$

What do you notice about the first and last digits of your answers?

3 (a) Copy and complete $99 \times 11 =$
$99 \times 22 =$
$99 \times 33 =$
$99 \times 44 =$
$99 \times 55 =$

(b) Write down what you think the answers to the following will be:

99×66, 99×77,
99×88, and 99×99.

(c) Check using your calculator and compare 99×11 and 99×99.

4 Copy and complete

(a) $15\,873 \times 7 =$
$15\,873 \times 14 =$

(b) What do you have to multiply 15 873 by to get 333 333?

(c) Try multiplying 15 873 by other multiples of 7 up to 63.

5^2 means $5 \times 5 = 25$
6^2 means $6 \times 6 = 36$

5 Copy and complete:

(a) $7^2 = \_\_ \times \_\_ = 49$

(b) $17^2 = \_\_ \times \_\_ = \_\_$

(c) $97^2 = \_\_ \times \_\_ = \_\_$

(d) $81^2 = \_\_ \times \_\_ = \_\_$

6 Copy and complete:

1^2	2^2	3^2	4^2	5^2	6^2	7^2	8^2	9^2	10^2	11^2	12^2	13^2	14^2	15^2
1	4	9	16											

Magic squares

7 (a) Add the numbers in each column and in each row.
(b) Add the numbers along the two diagonals.
What do you notice about your answers?

1	42	29	7	36	35
48	9	20	44	13	16
5	38	33	3	40	31
43	14	15	49	8	21
6	37	34	2	41	30
47	10	19	45	12	17

8 Copy this magic square and calculate the missing digits.

25		9	16	23
31	13	15	22	24
	14	21	28	
18	20	27		11
19	26		10	17

A multiplication square

1 Copy and complete this multiplication table on squared paper.

Use your completed table to answer questions 2, 3, 4, and 5.

X	1	2	3	4	5	6	7	8	9
1									
2								16	
3			9		15				
4				16	20	24			
5			15		25				45
6		12							
7					35				63
8							56		
9									

2

4	6	8	10	12
6	**9**	12	**15**	18
8	12	16	20	24
10	**15**	20	**25**	30
12	18	24	30	36

This '3 × 3' square is part of your multiplication table.

(a) Multiply the numbers in the opposite corners:

$9 \times 25 =$ $\quad$ $15 \times 15 =$

(b) What do you notice about your answers?

3 (a) Multiply the numbers in the opposite corners of the following squares (all taken from your table).

(b) What do you notice about your answers in each case?

4	8	12	16	20
5	**10**	15	**20**	25
6	12	18	24	30
7	**14**	21	**28**	35
8	16	24	32	40

4	5	6	7	8	9
8	**10**	12	14	**16**	18
12	15	18	21	24	27
16	20	24	28	32	36
20	**25**	30	35	**40**	45
24	30	36	42	48	54

4	6	8	10	12	14	16
6	**9**	12	15	18	**21**	24
8	12	16	20	24	28	32
10	15	20	25	30	35	40
12	18	24	30	36	42	48
14	**21**	28	35	42	**49**	56
16	24	32	40	48	56	64

4

6	8	10	12	14
9	12	**15**	18	21
12	**16**	**20**	**24**	28
15	20	**25**	30	35
18	24	30	36	42

(a) This '3 × 3' cross is part of your multiplication table. Outline this on your table.

(b) Add the numbers along each leg of the cross:

$15 + 20 + 25 =$ $\quad$ $16 + 20 + 24 =$

(c) What do you notice about your answers?

(d) What is the relationship between your answers and the number in the middle of the cross?

5

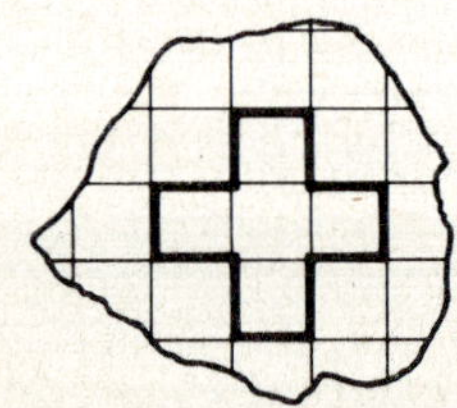

(a) Outline a **different** '3 × 3' cross on your table and add the numbers along each leg of this cross.

(b) What relationship is there between your answers and the number in the middle of the cross?

Shapes from triangles

Making the triangles

You need FOUR CARDBOARD TRIANGLES of the size shown.
Two possible ways of making them are given below.

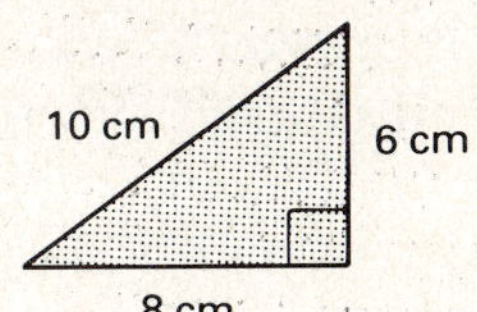

1 Draw **two** rectangles like this, measuring 8 cm by 6 cm, on centimetre squared paper.

Stick the two rectangles on cardboard. Cut out the **four** triangles.

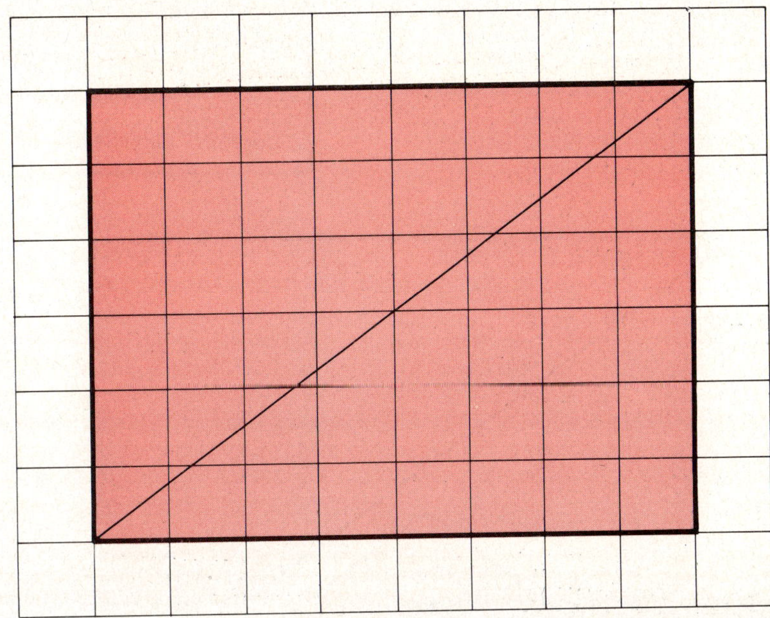

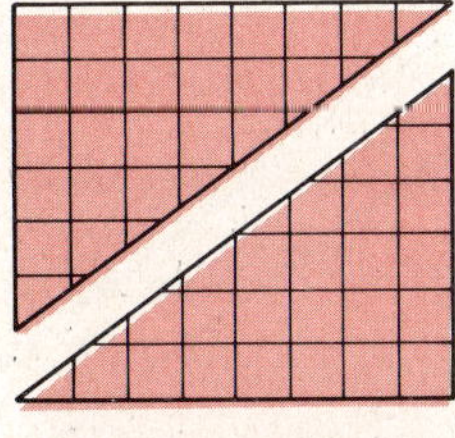

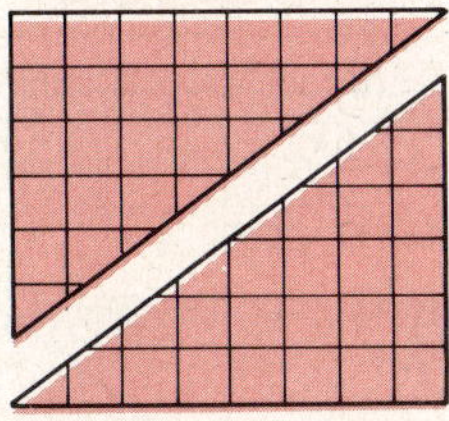

OR

2 Use a ruler to measure 8 cm and 6 cm in the square corner of a sheet of cardboard, as shown.

Draw a line and cut off a triangle.

Make **four** triangles like this.

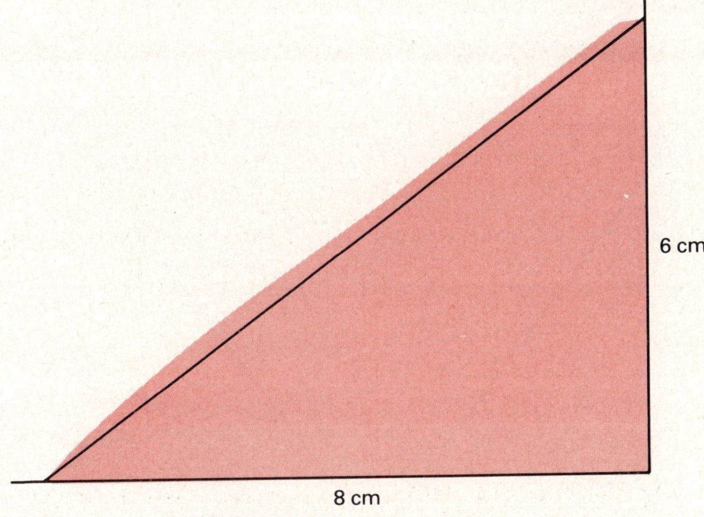

Keep the four triangles in an envelope.
You will need them for all the pages of this topic.

Making shapes with two triangles

1 Use **two** of your triangles to make each shape.

2 Draw round the pieces like this

to show how you made each shape.

3 Name each shape.

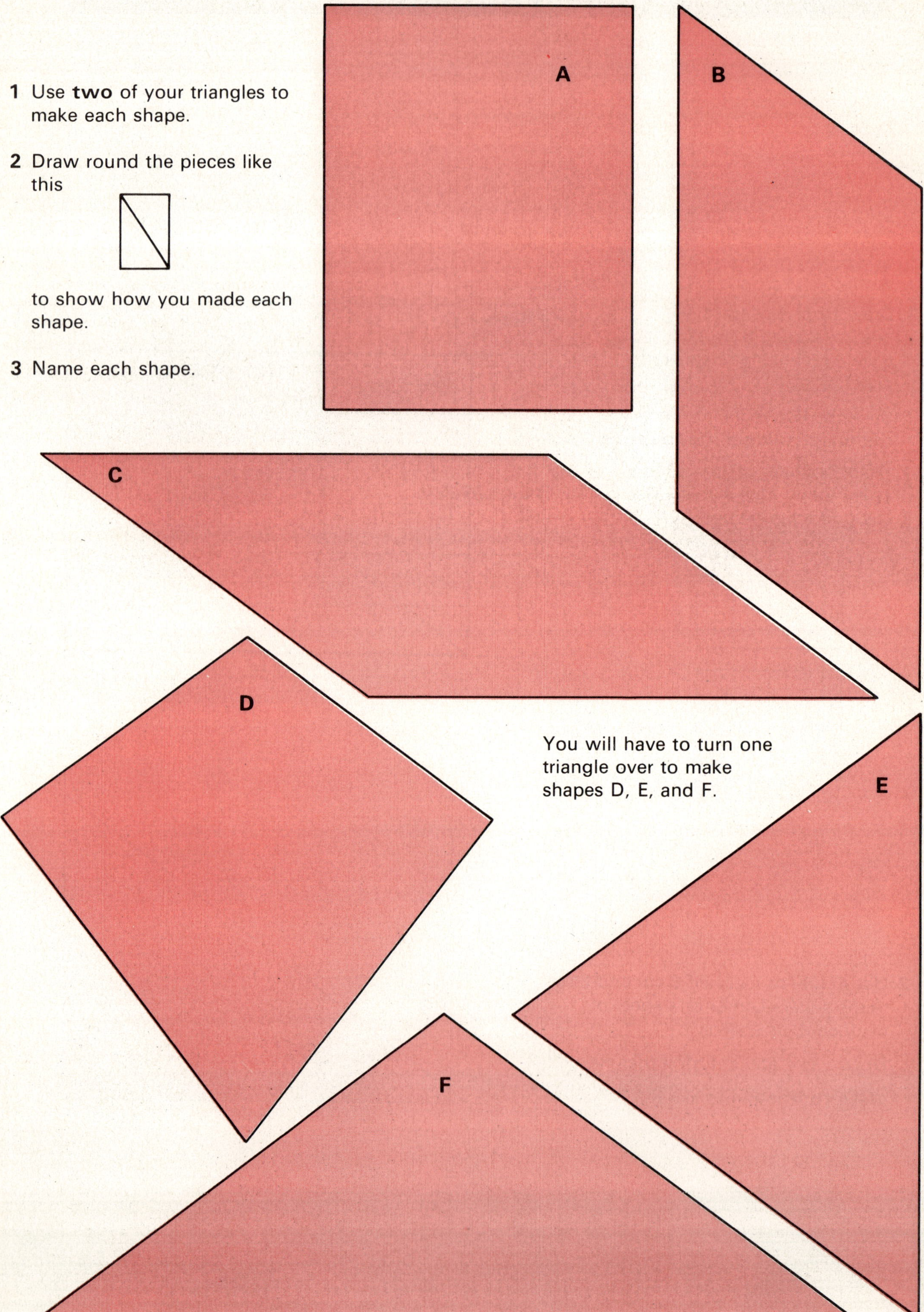

You will have to turn one triangle over to make shapes D, E, and F.

Perimeters and areas

Each shape on page 12 is made from TWO of these triangles.

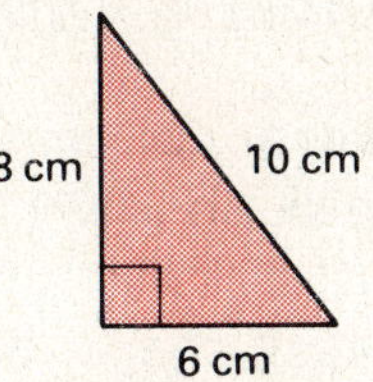

1 The **perimeter** of rectangle A = 8 + 6 + 8 + 6 cm
= 28 cm

(a) Copy the table below.
(b) Find the perimeters of the other shapes on page 12.
Use a ruler if you wish.

Fill in the 'Perimeter' column of your table.

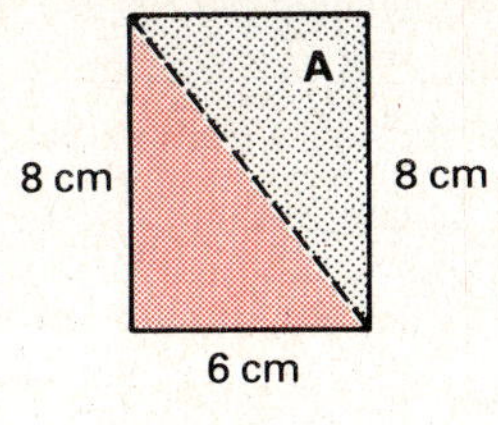

Shape	*Perimeter*	*Area*
A	28 cm	
B		
C		
D		
E		
F		

2 The **area** of rectangle A = 8×6 cm^2
= 48 cm^2

All the other shapes on page 12 are made from the same two triangles.
Fill in the 'Area' column of your table.

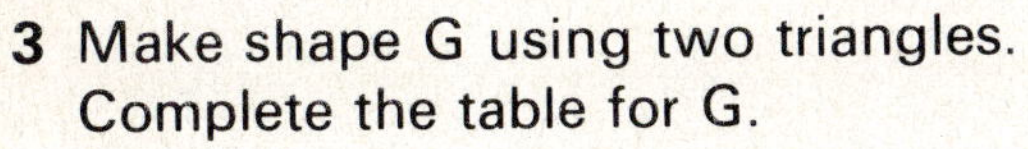

3 Make shape G using two triangles.
Complete the table for G.

4 Which of the statements below is true?
The answers in your table should help you to decide.
(a) Shapes with the same area can have different perimeters.
or
(b) Shapes with the same area must have the same perimeter.

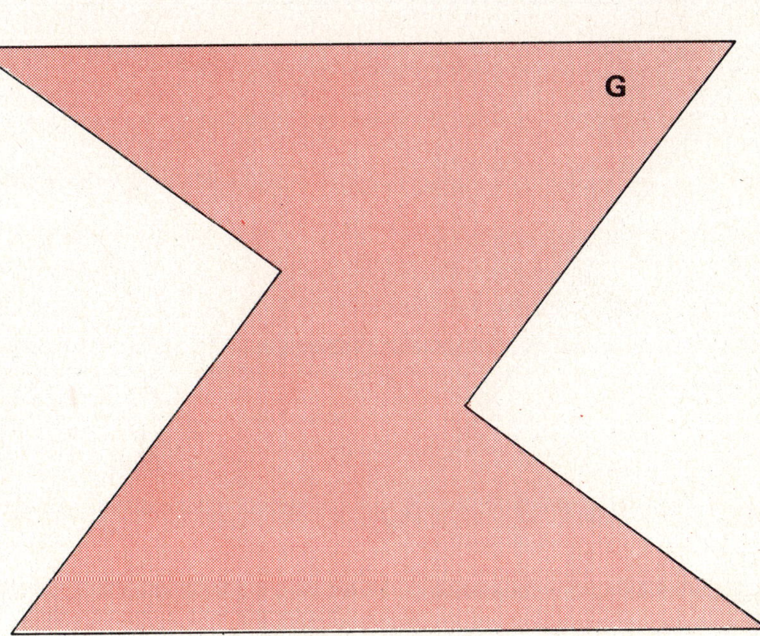

Making shapes with four triangles

1 Use all **four** triangles to make each shape on pages 14 and 15.

2 Show how you made each shape by drawing round the pieces **or** by making a small sketch.

3 Name each shape. Shape J is already named.

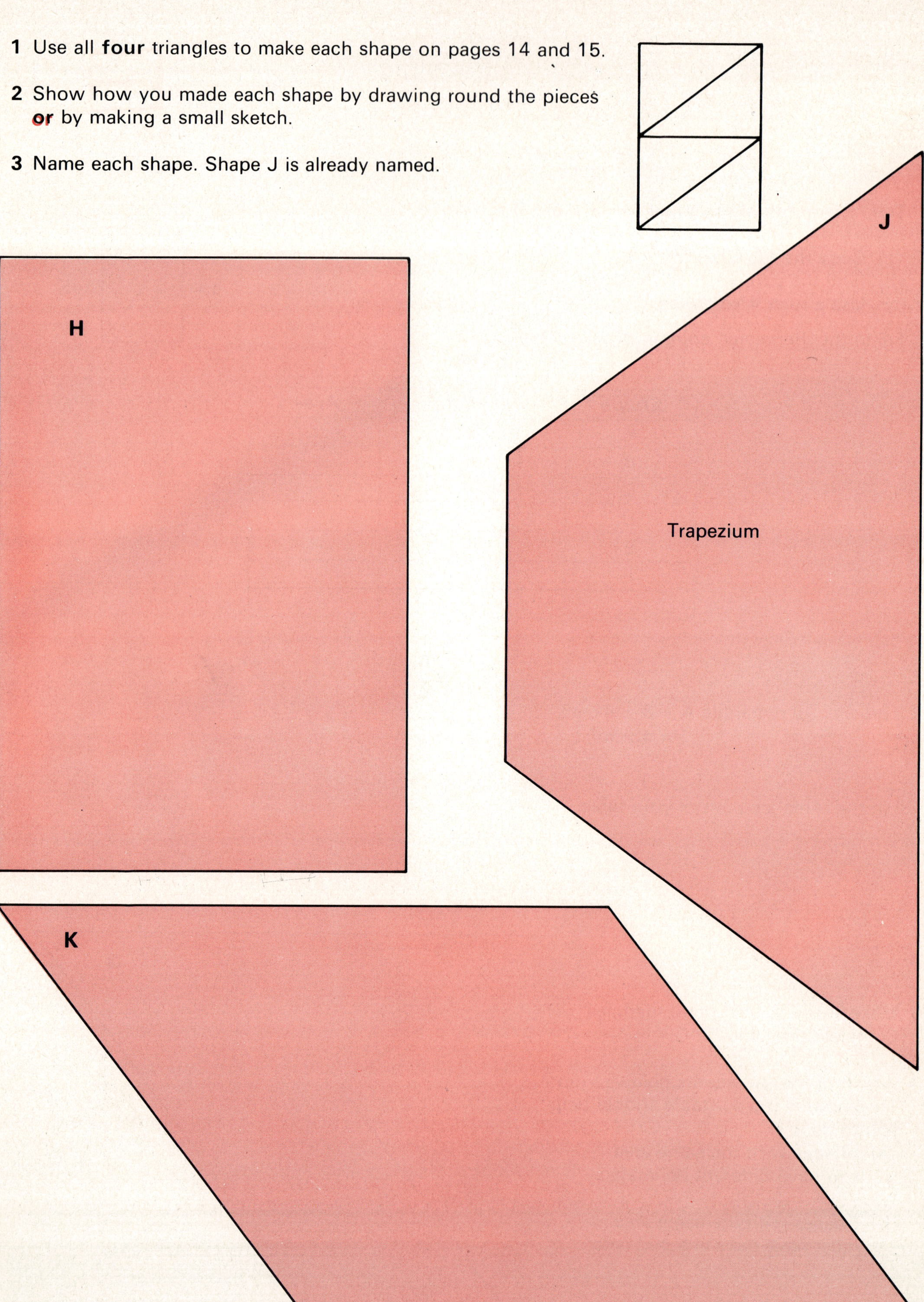

Making shapes with four triangles

Perimeters and areas

1 Use a ruler to find the perimeters, in centimetres, of pentagon P and square S. Write down your answers.

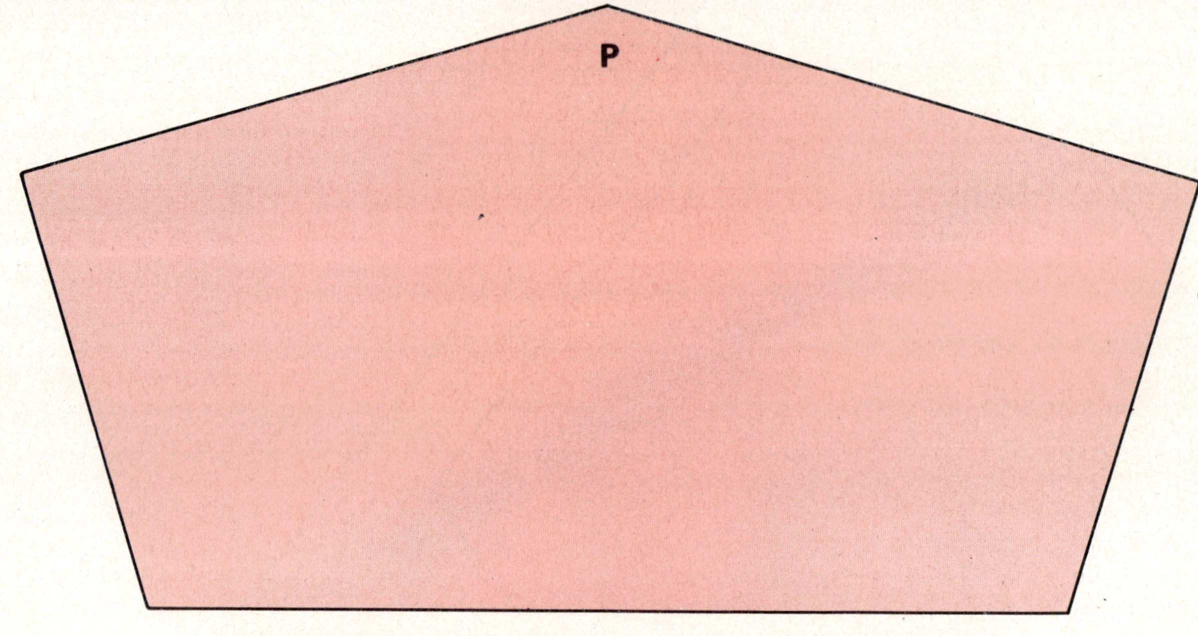

2 (a) Use the four cardboard triangles to make the pentagon P.
(b) Now use the same four triangles to make square S **but do not cover the black part**.
(c) Write down which shape, P or S, has the larger area.

3 Copy and complete this statement correctly:

Shapes with equal perimeters { **must have equal areas** / **or** / **can have different areas.** }

Drawing designs with triangles

1 Draw two lines at right angles in the middle of a sheet of paper. Make each line at least 20 cm long.

You could use **any one** of these methods:

Use squared paper

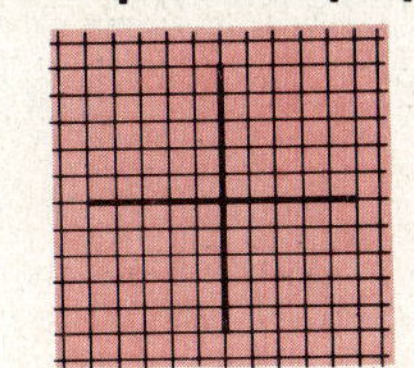

Fold a sheet of paper twice

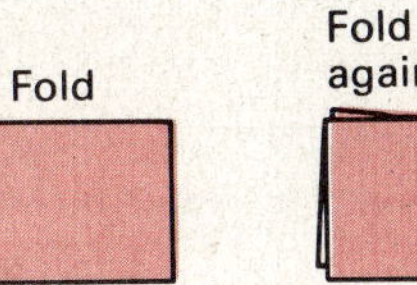

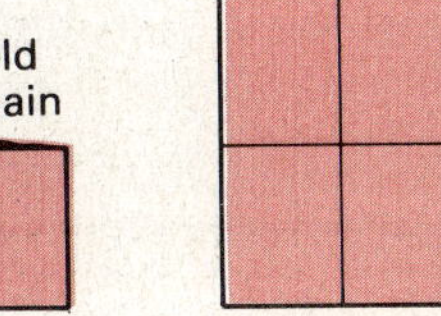

Use a set square

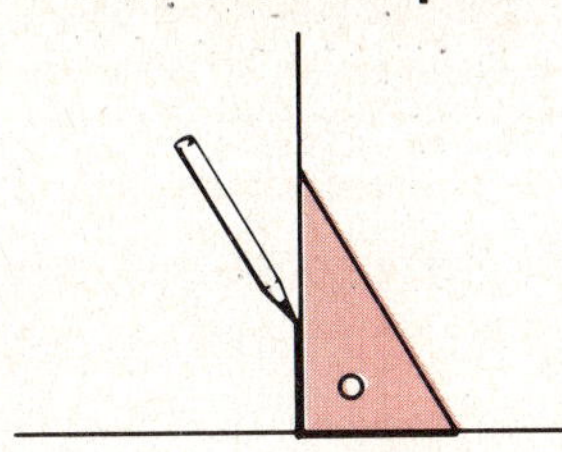

Use a protractor

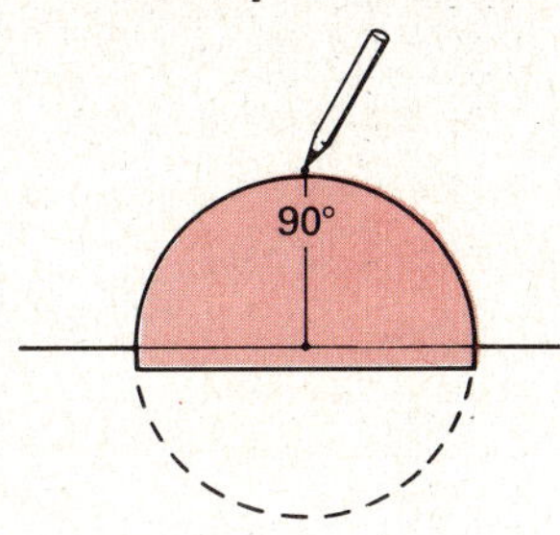

2 Starting with the two lines at right angles, draw the design below.
Draw round your triangle pieces to make the pattern.

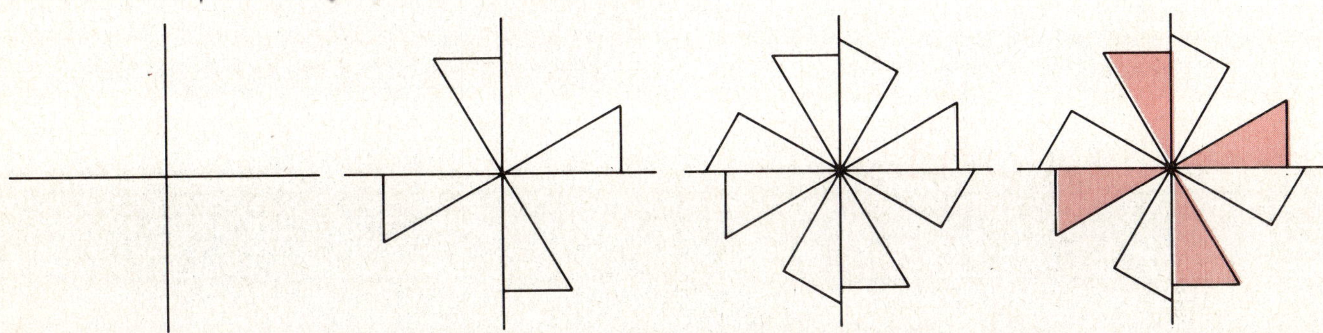

3 Try making one or two more designs by drawing round the triangles.
Each design below starts with two lines at right angles.

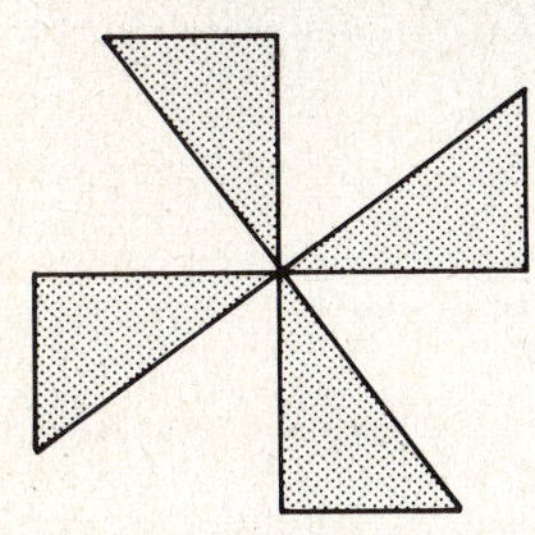

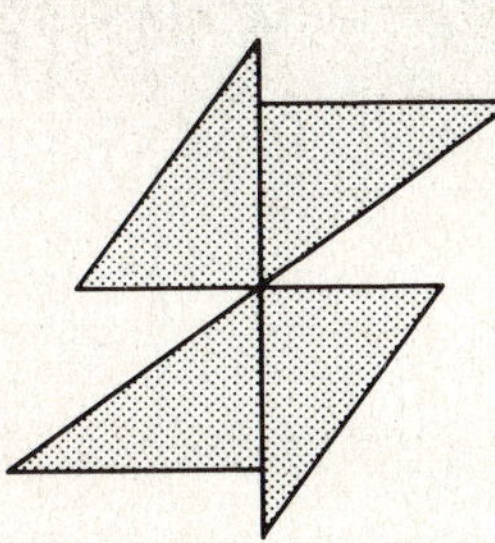

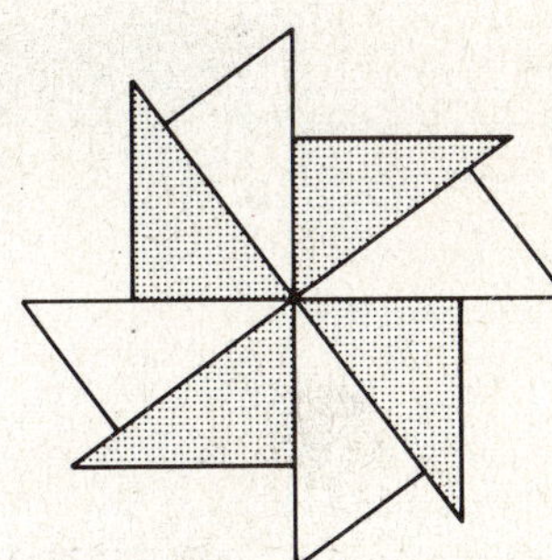

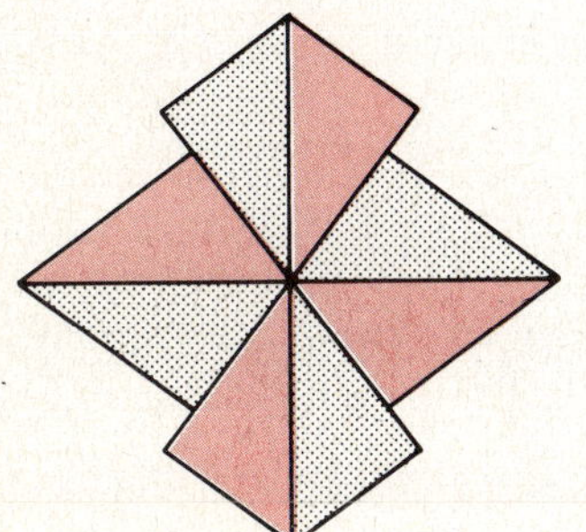

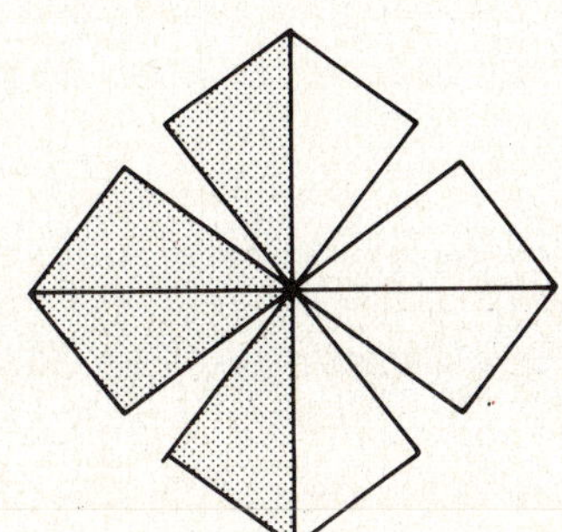

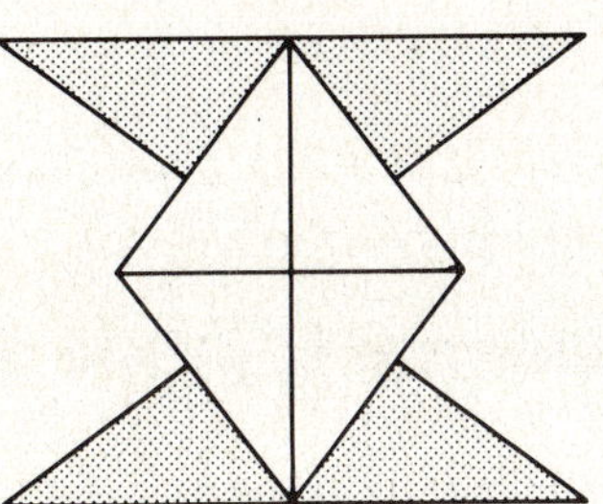

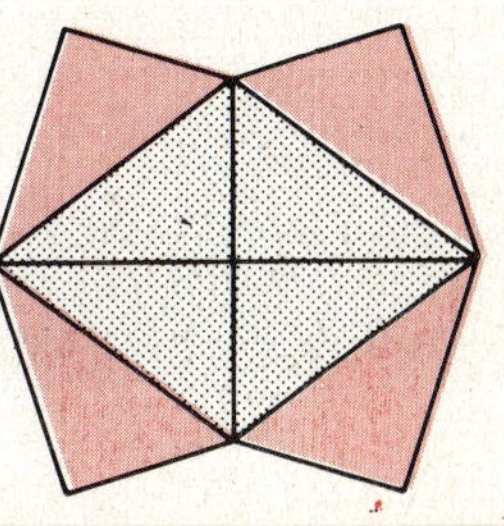

More designs

Try making one or two designs of your own.

You could use several sets of cardboard triangles and could fit them together to make a design.

You could then draw round the triangles to make your design and colour it.

Here are a few patterns to give you ideas. They do **not** start with two lines at right angles.

Distance graph

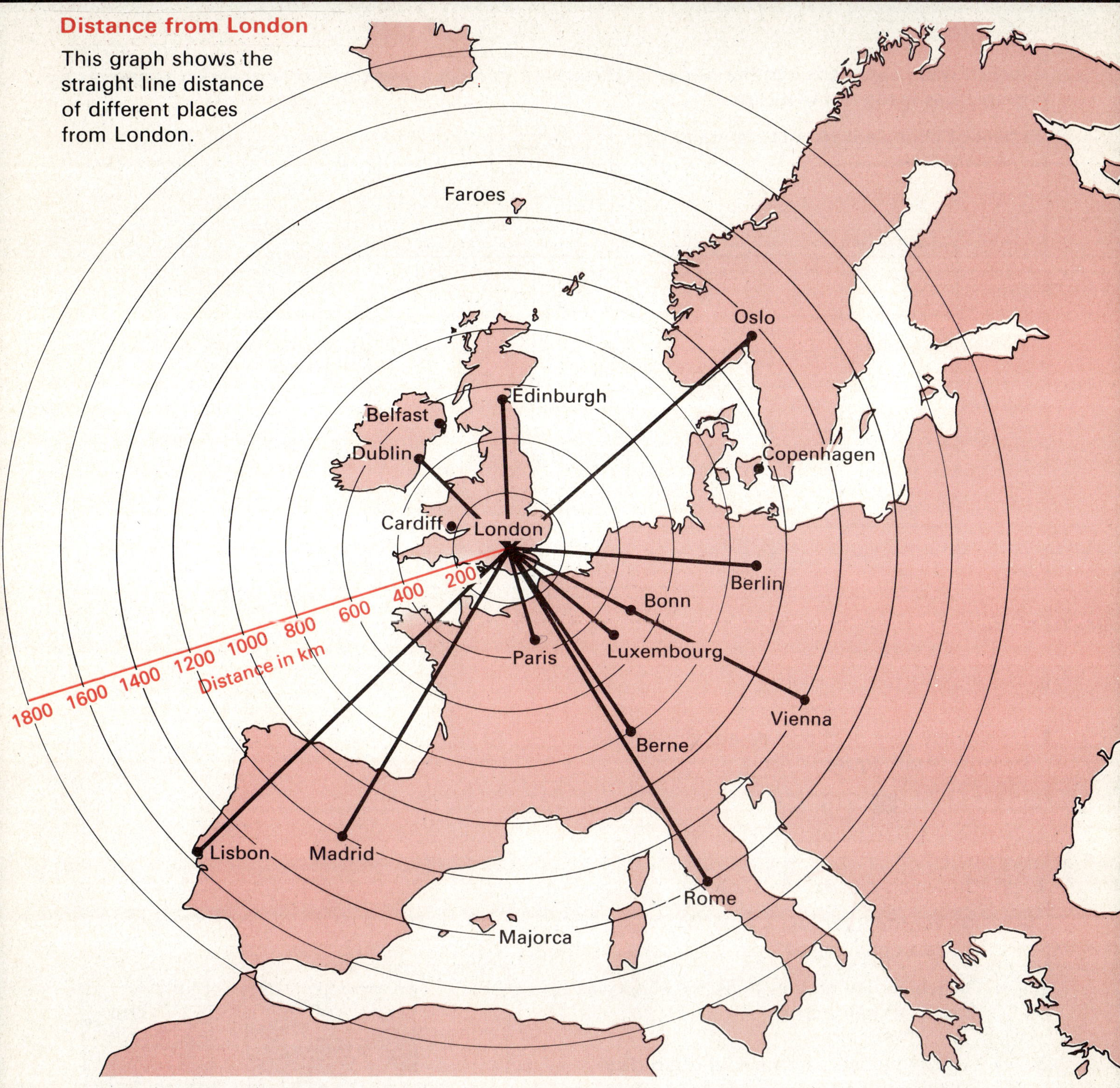

1 Which is nearer to London – Edinburgh or Paris?

2 How far are the following places from London?
(a) Berne (b) Madrid (c) Rome (d) Vienna

3 **Estimate** how far each of the following places is from London.
(a) Berlin (b) Luxembourg (c) Dublin (d) Cardiff

4 **Estimate** how far the following islands are from London.
(a) Faroes (b) Majorca

5 **About how far** is it from Lisbon to Oslo?

6 Most of these cities are the capitals of countries.
List capitals and countries like this:
Rome – Italy

Sale of chocolate

A record was kept of the number of bars of chocolate sold each week in a school tuck-shop. A graph was made of the results.

20 bars of chocolate were represented by: 

Chocolate graph

Day
Monday
Tuesday
Wednesday
Thursday
Friday

1 How many bars of chocolate are represented by **one small square**?

2 (a) On which day did the shop sell:

(i) most chocolate
(ii) least chocolate?

(b) On which day do you think the shop was closed for half a day?

3 (a) List the number of bars sold **each day**.
(b) How many bars were sold **altogether**?

4 How many more bars were sold on Tuesday than on Monday?

5 Here is a list of the number of bars of chocolate sold during the **next** week.

Monday	64 bars
Tuesday	78 bars
Wednesday	26 bars
Thursday	52 bars
Friday	87 bars

Use this information to make a graph like the one above.

Sale of cheese

Charlie Cheddar the grocer liked drawing graphs.

He kept a record of the number of boxes of cheese which he sold each week and made a graph of the results.

40 boxes sold were represented by 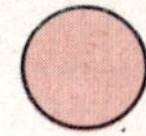

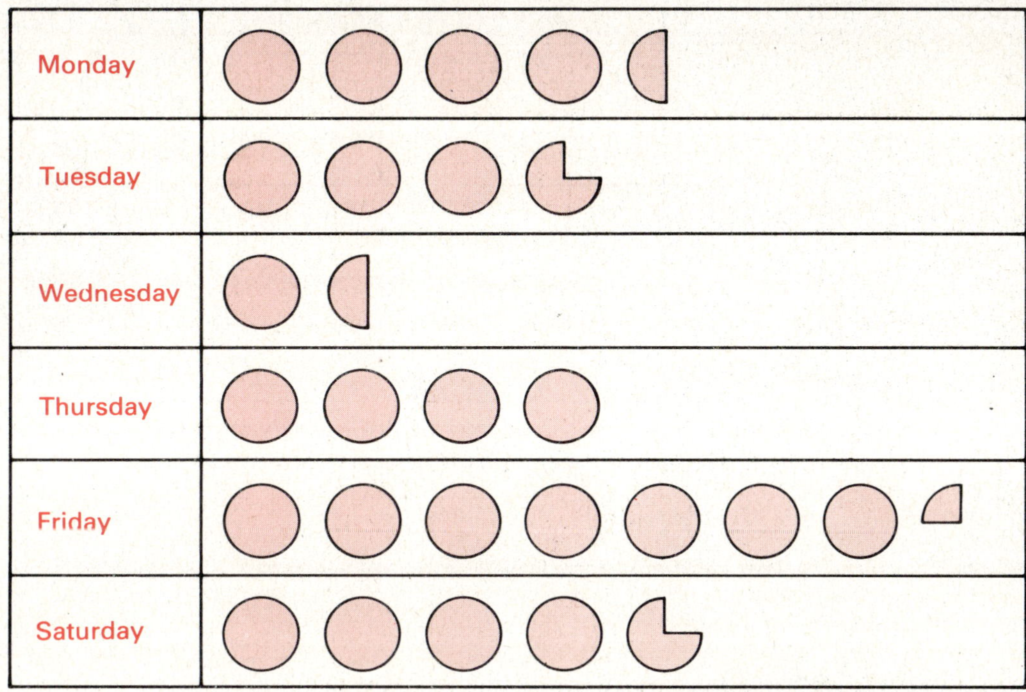

1 How many boxes of cheese are represented by:

(a) 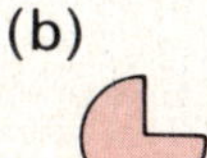 (b) 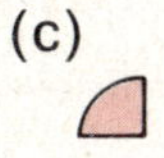 (c)

2 On which day did he sell:
(a) least cheese
(b) most cheese?

3 On which day do you think he:
(a) closed early
(b) stayed open late?

4 (a) List the number of boxes of cheese he sold **each day**.
(b) How many boxes did he sell altogether in the **week**?

5 How many more boxes did he sell on Monday than on Tuesday?

6 Here are the sales figures for another week.

Monday	160 boxes	Tuesday	180 boxes	Wednesday	90 boxes
Thursday	150 boxes	Friday	270 boxes	Saturday	170 boxes

Use this information to make a graph like the one above.

An addition square

1 Copy and complete this addition table on squared paper.

Use your completed table to answer questions 2, 3, 4, and 5.

+	1	2	3	4	5	6	7	8	9
1									
2							9		
3			6	7					
4			7	8	9				
5				9			12		14
6									
7							14		16
8	9								
9									

2

3	4	5	6
5	**6**	**7**	8
6	**7**	**8**	9
7	8	9	10

This '2 × 2' square is part of your addition table.
Multiply the numbers in the opposite corners.

$6 \times 8 =$ $7 \times 7 =$

The difference between your answers is 1.

3 Try this again for these '2 × 2' squares taken from your table.

6	7	8	9
7	**8**	**9**	10
8	**9**	**10**	11
9	10	11	12

$8 \times 10 =$
$9 \times 9 =$

11	12	13	14
12	**13**	**14**	15
13	**14**	**15**	16
14	15	16	17

$13 \times 15 =$
$14 \times 14 =$

8	9	10	11
9	**10**	**11**	12
10	**11**	**12**	13
11	12	13	14

$10 \times 12 =$
$11 \times 11 =$

Is the difference between your answers always 1?

4 Try this again for these '3 × 3' squares taken from your table.

4	5	6	7	8
5	**6**	7	**8**	9
6	7	8	9	10
7	**8**	9	**10**	11
8	9	10	11	12

$6 \times 10 =$
$8 \times 8 =$

7	8	9	10	11
8	**9**	10	**11**	12
9	10	11	12	13
10	**11**	12	**13**	14
11	12	13	14	15

$9 \times 13 =$
$11 \times 11 =$

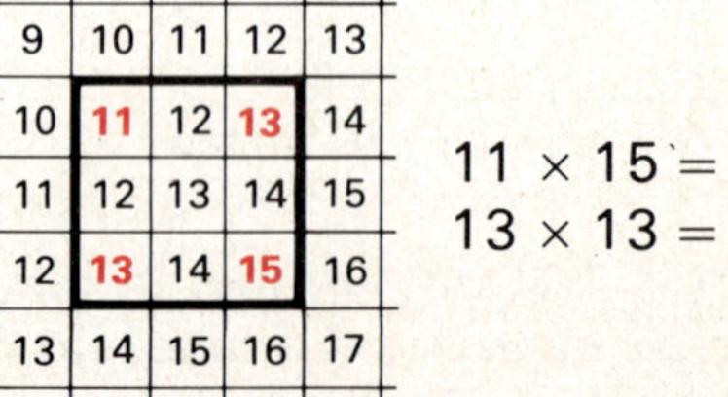

9	10	11	12	13
10	**11**	12	**13**	14
11	12	13	14	15
12	**13**	14	**15**	16
13	14	15	16	17

$11 \times 15 =$
$13 \times 13 =$

What is the difference between your answers this time?

5

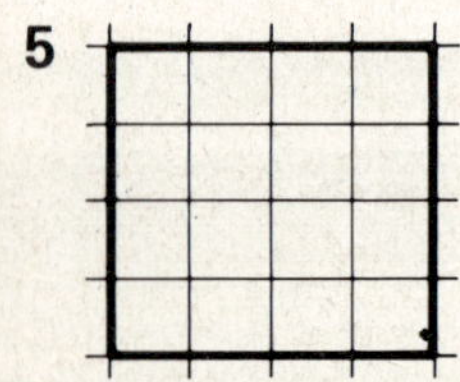

Outline a '4 × 4' square on your table.

(a) Multiply the numbers in the opposite corners.
(b) What is the difference between your answers this time?

What are the chances?

More likely, less likely

It is **more likely** to snow in Winter than in Summer.

It is **less likely** to snow in Summer than in Winter.

1 Copy and complete:

(a) In Winter it is **more likely** to be (cold/warm).

(b) In Spring you are **more likely** to see (daffodils/roses).

(c) In Summer you are **more likely** to play (indoors/outdoors).

(d) In Autumn you are **more likely** to pick (blackberries/raspberries).

2 Copy and complete:

(a) In a race between two boys the winner is **less likely** to be (the 11-year-old/the 6-year-old).

(b) A boy is **less likely** to take a penalty kick with his (right foot/left foot).

(c) A girl is **less likely** to play tennis (right handed/left handed).

(d) It is **less likely** that a girl will play (football/hockey).

3 Here is some information about the children in the circle.

Name	*Height*	*Brothers or sisters*	*Wears glasses*
David	152 cm	1 sister	No
Graham	148 cm	2 brothers	Yes
John	143 cm	1 sister 1 brother	No
Ian	155 cm	None	No
Moira	144 cm	None	No
Elizabeth	146 cm	2 sisters	No
George	151 cm	1 brother	No

Is Tom **more likely** to point to:

(a) a boy/a girl?

(b) a person who wears glasses/a person who does not wear glasses?

(c) a person whose name has 4 letters or less/a person whose name has more than 4 letters?

(d) a person with brothers or sisters/an only child?

(e) a person over 150 cm/a person less than 150 cm?

Two ways

Work with a partner. You will need a drawing pin and a nut.

When a drawing pin is dropped on a flat surface, it settles in one of these two ways.

point up or point down

1 Which way do you **think** is **more** likely?

2 (a) Drop a drawing pin 20 times.
Your partner can record the results like this:

point up	✓ ✓ ✓ ✓
point down	✓ ✓

(b) How many times did it settle (i) point up (ii) point down?
(c) Now ask your partner to drop the pin 20 times. You record the results.
(d) This time, how many times did it settle (i) point up (ii) point down?
(e) Were your results about the same as your partner's?

3 If you dropped 100 similar drawing pins, do you **think more** would settle 'point up' or 'point down'?

When a hexagonal nut is dropped on to a flat surface, it settles in one of these two ways.

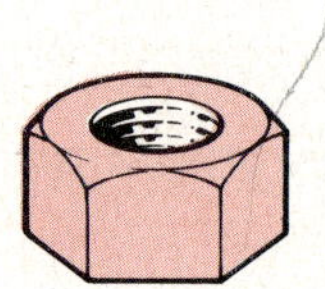

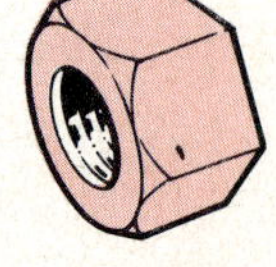

flat or on edge

4 Which way do you **think** is **less** likely?

5 (a) Drop a nut 20 times.
Your partner can record the results like this:

flat	✓ ✓ ✓ ✓
on edge	✓

(b) How many times did it settle (i) flat (ii) on edge?
(c) Now ask your partner to drop the nut 20 times. You record the results.
(d) How many times did it settle (i) flat (ii) on edge this time?
(e) Were your results about the same as your partner's?

6 If a box containing 200 of these nuts is emptied on to a table top, do you **think more** would settle 'flat' or 'on edge'?

Predicting

Work with a partner.
You will need a pack of 52 playing cards and some coloured beads.

1 (a) Count the number of picture cards in the pack of cards.
(b) If you shuffle the cards and choose one by cutting the pack, are you **more likely** to choose a picture card or a non-picture card?

2 (a) Cut the shuffled pack 20 times and see if you have a picture card or a non-picture card.
Your partner can record the results like this:

P, N, N, N, N, P . . . . .

(Write 'P' for picture card, 'N' for non-picture card.)

(b) How many times did you cut (i) a picture card
(ii) a non-picture card?

(c) Does your answer agree with your answer for question 1 (b)?

Put 2 red beads and 1 blue bead in a bag.

3 If you shake the bag behind your back and pick out one bead, which colour are you **more likely** to pick out?

4 (a) Now pick out a bead 15 times.
(Remember to put the bead back in the bag each time.)
Your partner can record the results like this:

R, R, B, B, B, ...

(Write 'R' for a red bead, and 'B' for a blue bead.)

(b) Now ask your partner to pick out the bead 15 times. You record the results.

(c) Did you both pick out a red bead about the same number of times?

(d) Did you get the answer you predicted in question 3?

5 If you pick a bead out of the bag 100 times (replacing the bead each time), which colour do you **think** you would pick most often?

Most likely, least likely

A team at the top of a league plays a team at the bottom of that league.

There are three possible results.

(1) The top team will win.
(2) It will be a draw.
(3) The bottom team will win.

It is **most likely** that the top team will win.
It is **least likely** that the bottom team will win.

1 Today a packet of crisps costs 10p.
A year from now the cost could still be 10p, could be more than 10p, or could be less than 10p.
Which do you think is (a) most likely (b) least likely?

2 At traffic lights on a certain road, red shows for 20 seconds, amber for 5 seconds, and green for 15 seconds.
When you arrive at these lights in a bus, which colour is
(a) most likely (b) least likely to be showing?

3 A bag holds 8 beads (1 red, 2 blue, and 5 white).
The bag is shaken and, without looking, a boy picks out one bead.
Which colour of bead is (a) most likely (b) least likely to be picked out?

4 A pack of playing cards contains 4 aces, 12 picture cards, and 36 number cards.
If the pack is shuffled and one card chosen, which type of card (ace, picture, or number) is
(a) most likely (b) least likely to be chosen?

A bag holds **only** red beads. If you shake the bag and pick out one bead it must be a red bead.

We say that you are **certain** to pick a red bead.

There are no blue beads in the bag.

We say that it is **impossible** to pick a blue bead.

5 Copy and complete these sentences using either **certain** or **impossible**.
(a) If today is Wednesday, it is .... that tomorrow is Friday.
(b) If you throw an ordinary dice, it is .... that the number showing will be an 8.
(c) If you roll a ball it is .... that it will eventually stop.
(d) It is .... for you to watch B.B.C. and I.T.V. on the same television set at the same time.
(e) If a child is born on the 1st April, it is .... that he will have his birthday on the 1st April every year.

Three ways

Work with a partner.
You will need a yogurt carton and a pack of playing cards.

When an empty yogurt carton is dropped it can settle in these three ways.

on its side　　on its base　　upside down

1 Which way do you **think** is (a) most likely (b) least likely?

2 (a) Drop a yogurt carton 25 times. Your partner can record how it settled by completing a table like this:

		Total
on its side	~~||||~~ ||	
on its base	|||	
upside down	||	

(b) Do these results agree with your answers for question 1?
(c) Now ask your partner to drop the carton 25 times. You record the results.
(d) Did you and your partner get about the same answers?

3 If you dropped the carton 200 times, in which way do you think it would settle
(a) most often (b) least often?

When you cut a pack of playing cards you can get one of three types of card.

a picture card　　a number card　　an ace

4 (a) Count how many (i) picture cards (ii) number cards (iii) aces there are in a pack of cards.
(b) Which type of card are you (i) most likely (ii) least likely to cut?
Give a reason for your answer.

5 (a) Cut a pack of cards 25 times. Your partner can record the results by completing a table as shown:

		Total
picture card	||	
number card	~~||||~~		
ace			

(b) Did you predict correctly which type of card was (i) most likely (ii) least likely?
(c) Now ask your partner to cut the cards 25 times. You can record the results.
(d) Did you and your partner get about the same answers?

6 Do you think you are more likely, less likely, or equally likely to cut a picture card as an ace?
Give a reason for your answer.

Equally likely

Work with a partner.
You will need a coin and a dice.

When a coin is tossed is it **equally likely** to settle 'heads' as 'tails'?

1 (a) If you tossed a coin 10 times, about how many (i) heads (ii) tails would you expect?
(b) If you tossed a coin 100 times, about how many (i) heads (ii) tails would you expect?

2 (a) Toss a coin 10 times.
Your partner can record the results like this: H, H, T, H, T . . . . .
(Write 'H' for heads, and 'T' for tails.)
(b) How many times did you get (i) a head (ii) a tail?
(c) Now ask your partner to toss the coin 10 times. You record the results.
(d) How many times did your partner get (i) a head (ii) a tail?
(e) Did you both get a head about the same number of times?
(f) How many heads did you get for the total of 20 tosses?

If you throw a dice it can 'fall' in 6 different ways.

Three of the ways will show an **odd** number ⟶

Three of the ways will show an **even** number ⟶

3 (a) Do you think you are (i) more likely (ii) less likely (iii) equally likely to throw an odd or an even number?
(b) If you throw a dice 10 times, about how often would you expect to throw an odd number?
(c) For 100 throws, about how often would you expect to throw an even number?

4 (a) Throw a dice 10 times.
Your partner can record the results like this: O, O, E, O, E, ...
(Write 'O' for odd, and 'E' for even.)
(b) How many times did you get (i) an odd (ii) an even number?
(c) Now ask your partner to throw the dice 10 times. You record the results.
(d) How many times did your partner get (i) an odd (ii) an even number?
(e) Did you and your partner each get about the same number of 'odds'?
(f) How many 'odds' did you get altogether for the total of 20 throws?
How many did you expect? Why?

A code

Codes have been used throughout history to disguise the way we write information.

A code like this was used in the 16th century.

A B C	D E F	G H I
J K L	M N O	P Q R
S T U	V W X	Y Z

For A we write

For F we write

For H we write

For M we write

For Q we write

For Y we write

Can you see how the code works?

1 Write down the code for

(a) B (b) D (c) I (d) L
(e) N (f) T (g) X (h) Z

2 Decode the following messages

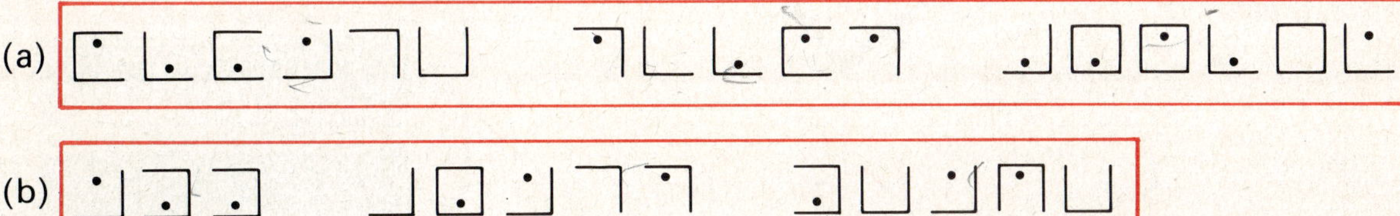

3 Write the following messages in code

(a) MEET ON SATURDAY
(b) JOHN KNOWS THE PASSWORD
(c) FIONA HAS THE KEY

4 Make up another message and write it in code.

Machines and money

Look at this receipt. It shows

(a) the date	19 MARCH 82
(b) the cost of each item	001.62
	000.28
(c) the total cost	001.90 TOTL
(d) the money handed to the shop assistant	005.00 AMT TEND
(e) the change.	003.10 CHNG

1 Look at this receipt.

(a) What is the date?
(b) What is the number of items bought?
(c) What is the total cost?
(d) How much money is handed to the shop assistant?
(e) How much change is given?

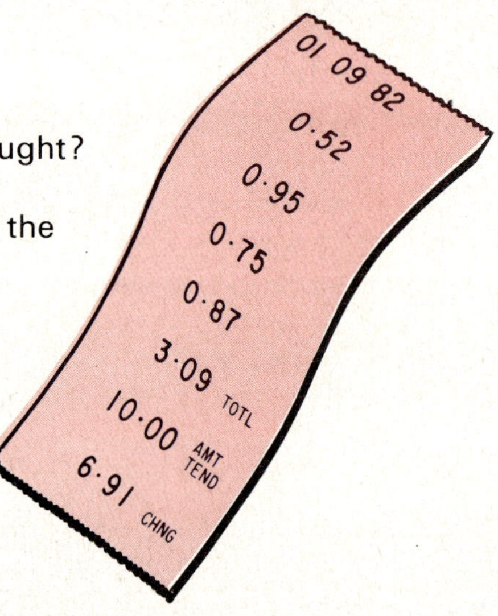

2 Collect receipts like these. Discuss with your teacher what each one shows.

This petrol pump shows how much one litre of petrol costs.

3 (a) What cost will show on this pump?

(b) What volume will show on this pump?

4 Find out what petrol pumps at your local garage look like. Discuss with your teacher what they show.

In a car park you can pay by putting 10p and 5p coins in a machine.
The machine shows how much is still to be paid.

5 (a) List the coins which were put in this machine.

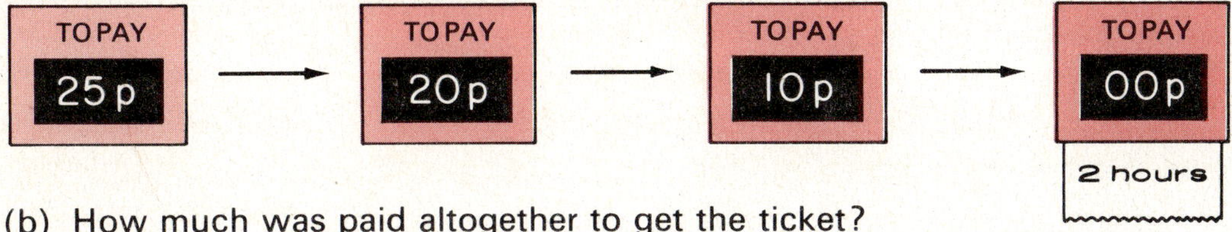

(b) How much was paid altogether to get the ticket?

6 (a) What amounts would show on this machine?

(b) How much was paid altogether to get this ticket?

7 Find out about machines in car parks and parking meters near where you live.
Discuss what they show with your teacher.

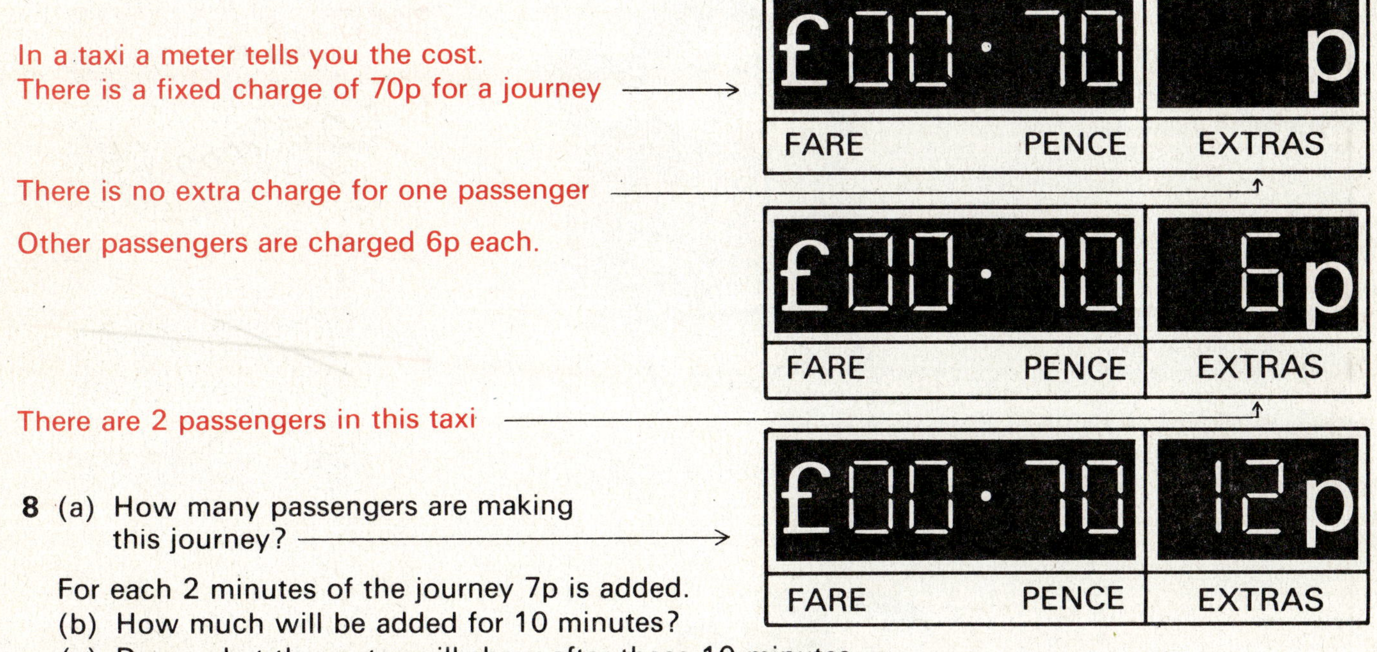

In a taxi a meter tells you the cost.
There is a fixed charge of 70p for a journey

There is no extra charge for one passenger

Other passengers are charged 6p each.

There are 2 passengers in this taxi

8 (a) How many passengers are making this journey?

For each 2 minutes of the journey 7p is added.

(b) How much will be added for 10 minutes?

(c) Draw what the meter will show after these 10 minutes.

(d) If the total cost is shared equally, how much does each passenger pay?

9 Find out about other machines which tell you about money.
Discuss what they show with your teacher.

Models from nets

Shapes you will need

For this topic you need a SQUARE and an EQUILATERAL TRIANGLE with the same edge length, say 5 cm.

If you do not have suitable plastic shapes, ask your teacher how you should make cardboard ones.

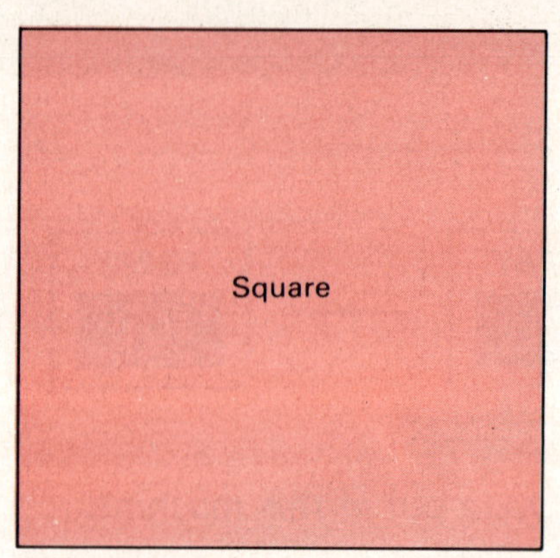

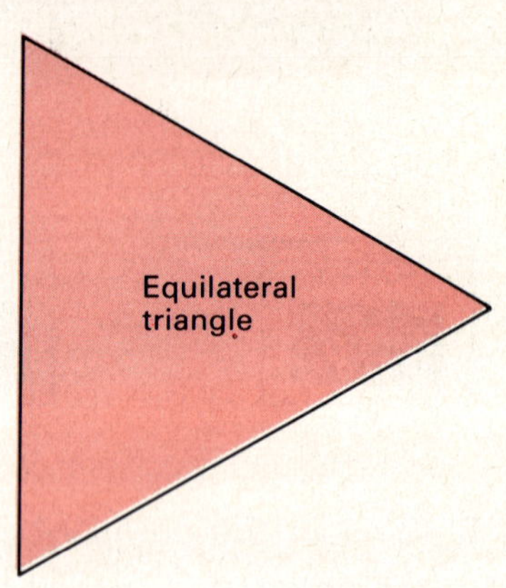

An umbrella

1 Use your triangle to draw this net on cardboard.

2 Cut out the net in one piece.
Now cut along the thick **coloured** line.

3 Lightly score along the dotted lines with the point of a pair of scissors.

4 Fold along the dotted lines. Tuck triangle 'A' under triangle 'B' and glue it to the underside of 'B'.

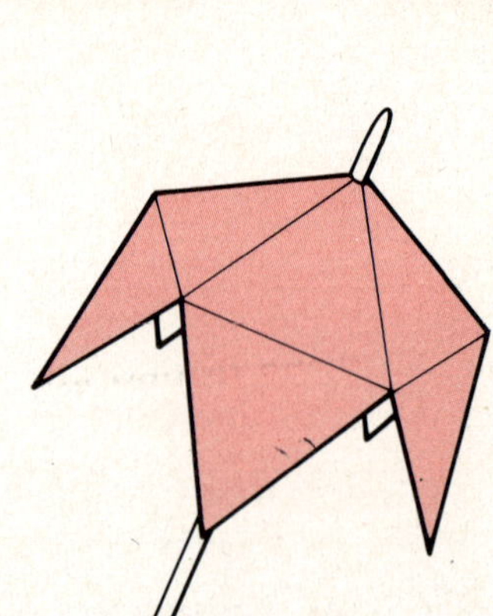

5 Push a round lollipop stick through the centre and glue it there to make the handle of the umbrella.

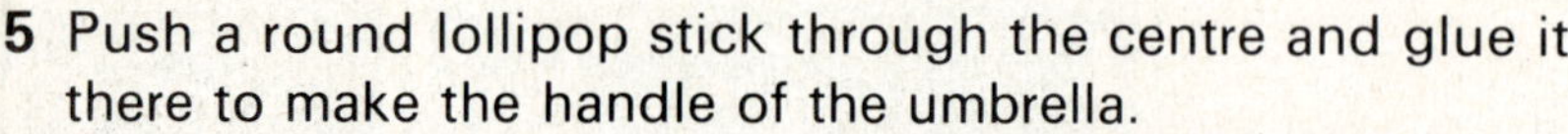

A bell flower

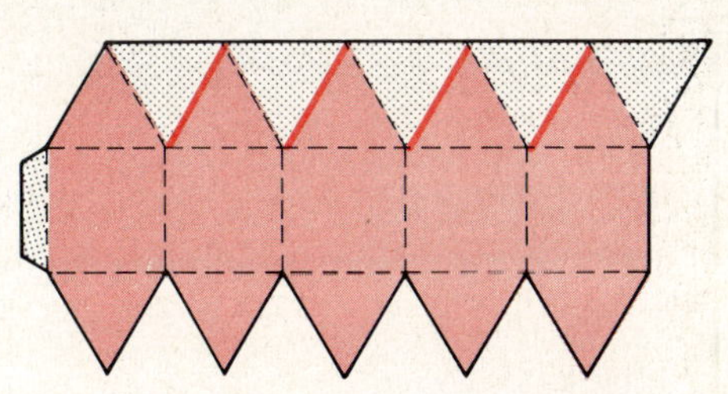

1 Use your square and triangle to draw this net on cardboard. The grey parts are flaps which will be glued later.

2 Cut out the net and flaps in one piece. Now cut along the thick **coloured** lines.

3 Score lightly along the dotted lines with the point of a pair of scissors.

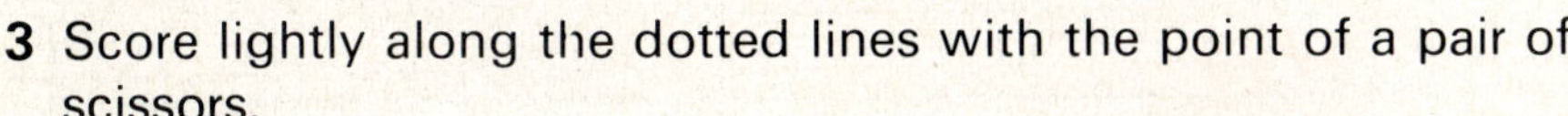

4 Fold along the dotted lines. Glue the flaps under the other triangles to form a pyramid shape. The pyramid becomes the base of the flower.

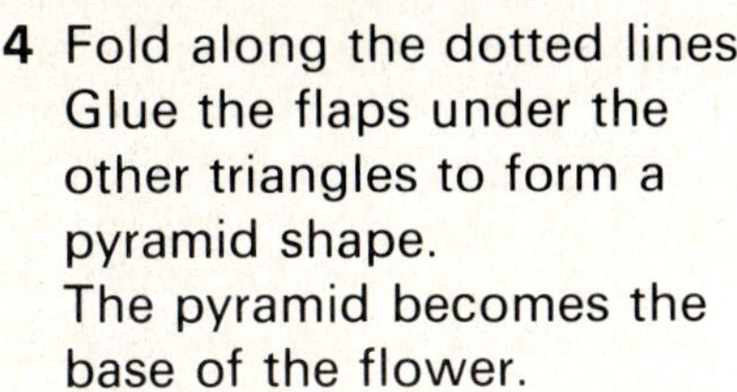

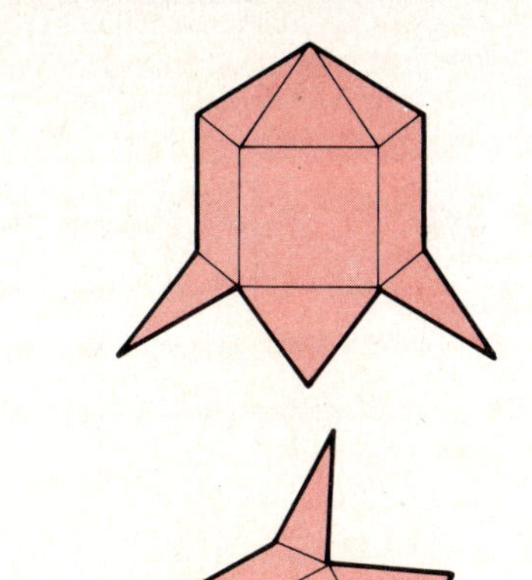

5 Push a pipe cleaner through the base and glue it there to make a stem.

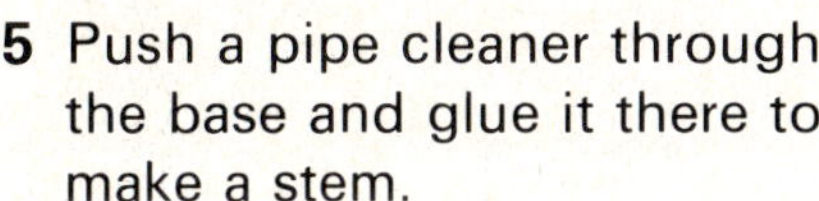

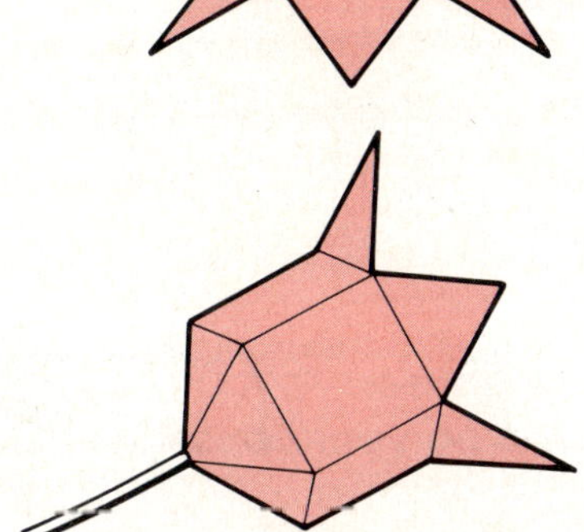

A gift box

1 On a piece of cardboard draw round your triangle **6 times** to make a hexagon like this. Rub out the dotted lines.

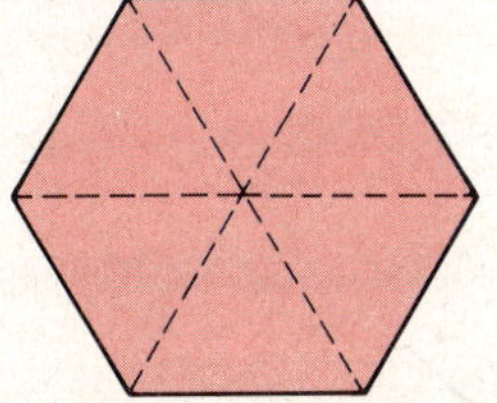

2 Now use your square and ruler to draw the rest of this net. ⟶ Cut out the net in one piece.

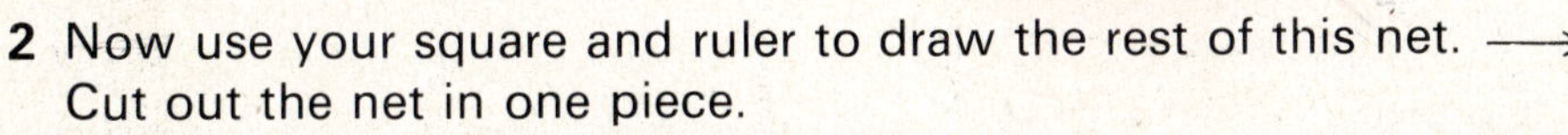

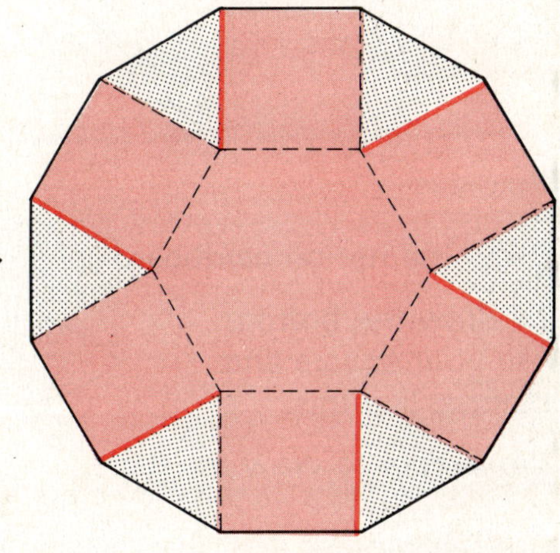

3 Cut along the thick **coloured** lines.

4 Score along all the dotted lines.

5 Fold up to make the box. Glue the flaps to the **inside**. Decorate your box.

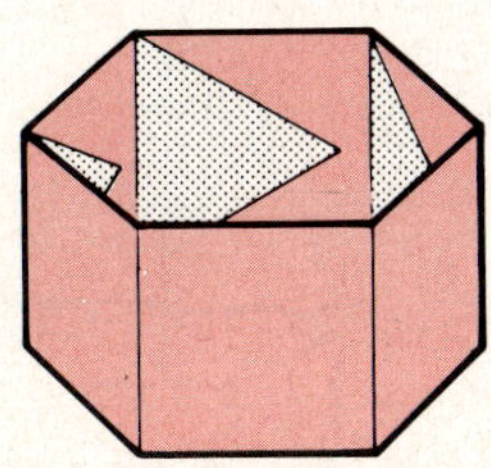

A house

This is the net for a house.

The grey parts are flaps which will be glued later.

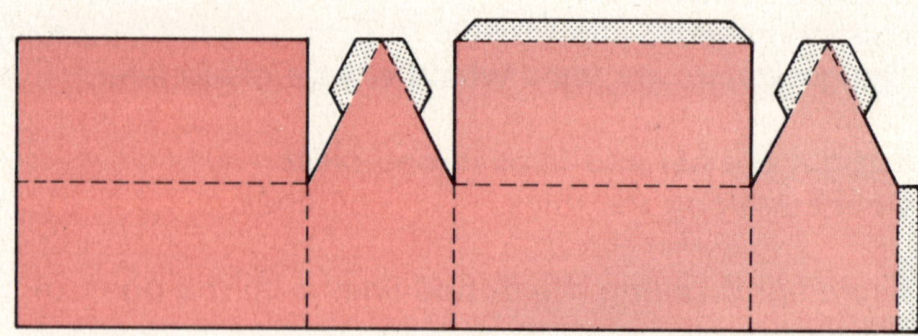

1 Draw the net on cardboard.

To draw the rectangular walls use your square **twice**.

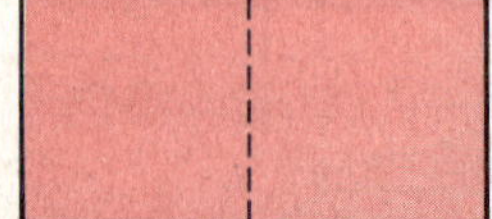

Remember to draw the flaps.

To draw the gable end use your square and triangle like this ⟶

2 Cut out the net with flaps in one piece.

3 Score lightly along the dotted lines.

4 Fold and glue the flaps inside to make the house.
You may want to paint doors and windows on your house.

A boat

This is the net for a boat.

The grey parts are flaps which will be glued later.

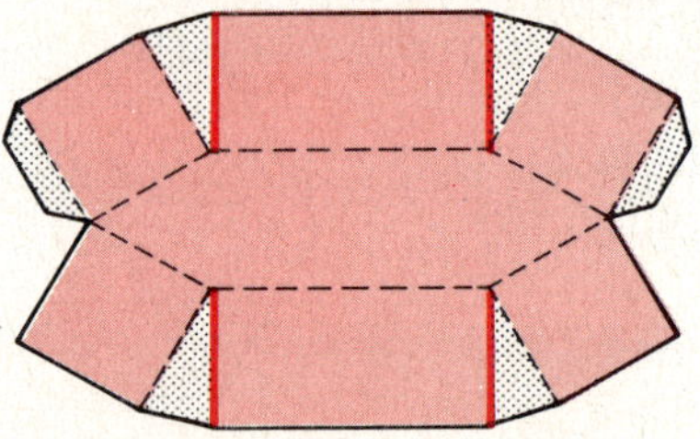

1 Draw the net on cardboard.

To draw the bottom use your square and triangle like this.

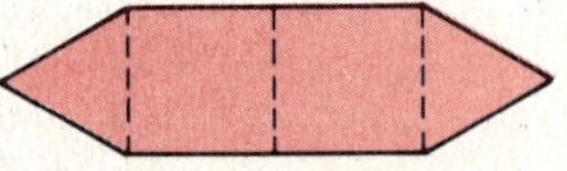

Remember to draw the flaps.

To draw the sides use your square twice like this.

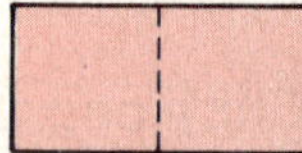

2 Cut out the net with flaps in one piece.

3 Now cut along the thick **coloured** lines.

4 Score lightly along the dotted lines.

5 Fold and glue the flaps inside to make the boat.

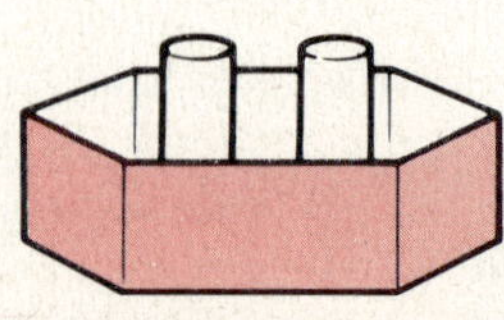

6 You could make paper funnels or paper sails with a drinking-straw mast.
Mast and funnels could be fitted to the bottom of the boat with plasticine.

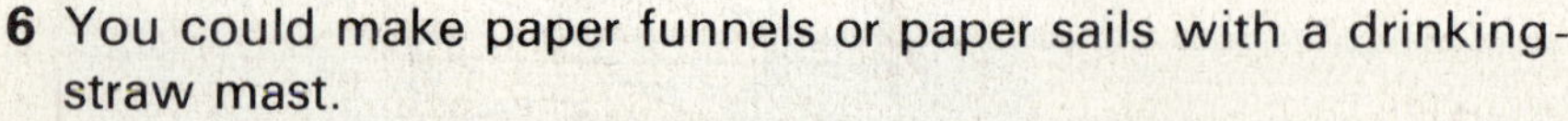

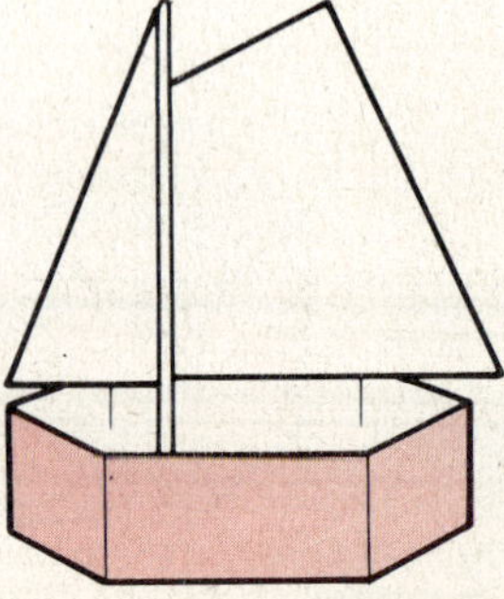

A spinning top

This net with flaps makes a pentagonal pyramid.

It can be used to make a spinning top.

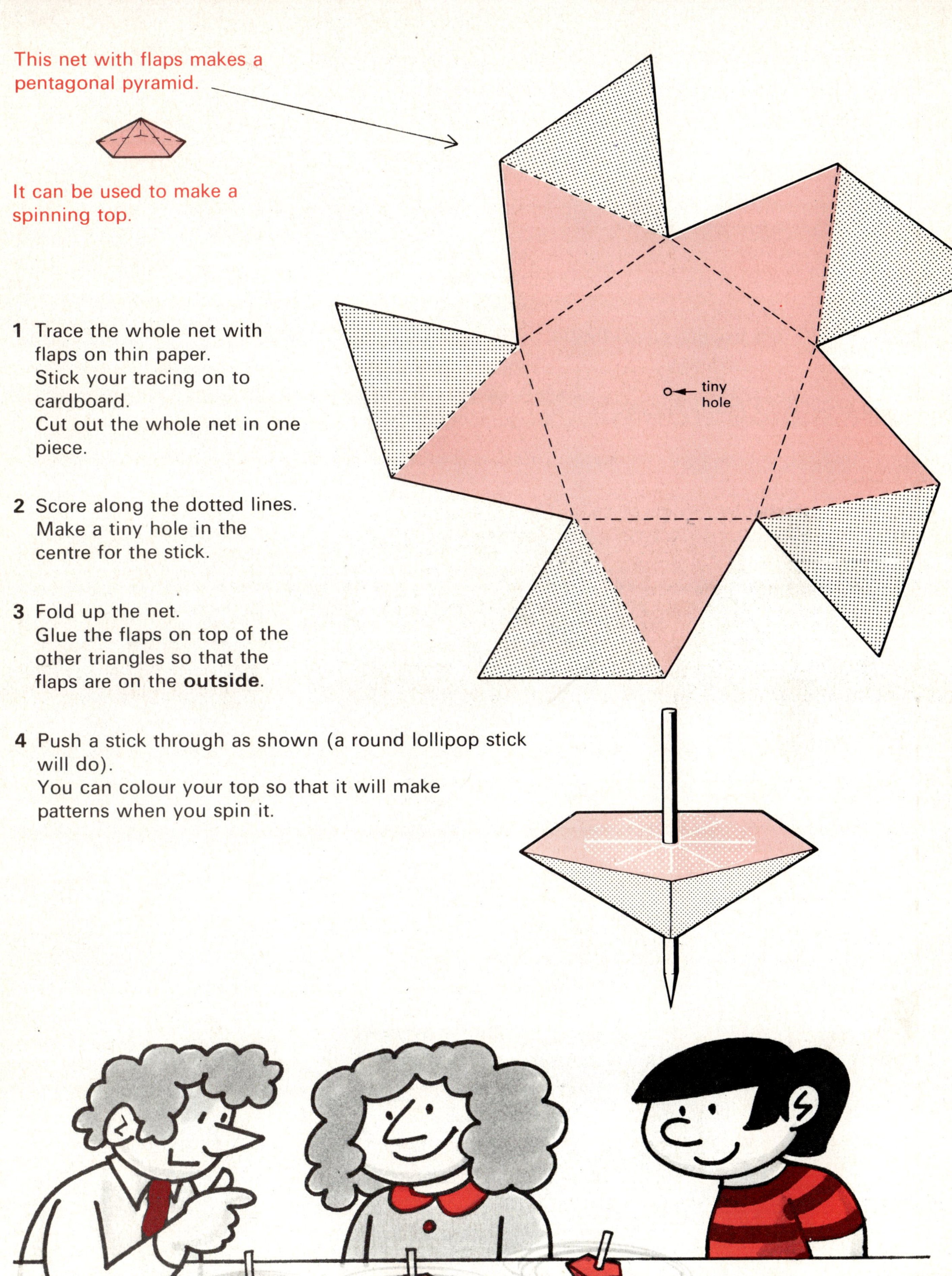

1 Trace the whole net with flaps on thin paper.
Stick your tracing on to cardboard.
Cut out the whole net in one piece.

2 Score along the dotted lines.
Make a tiny hole in the centre for the stick.

3 Fold up the net.
Glue the flaps on top of the other triangles so that the flaps are on the **outside**.

4 Push a stick through as shown (a round lollipop stick will do).
You can colour your top so that it will make patterns when you spin it.

A pyramid puzzle

This is the net with flaps of ONE of the pieces of the puzzle.

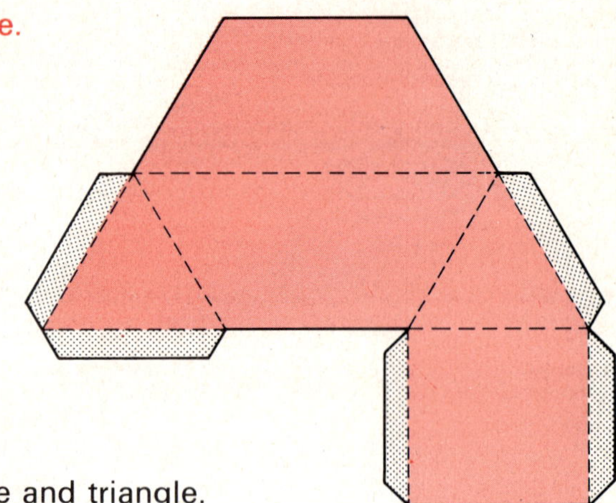

Its faces are:

two triangles  

one square 

two trapeziums 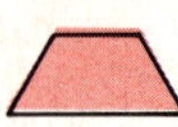 

1 Draw TWO nets like this on cardboard, using your square and triangle.

You could draw the trapezium like this

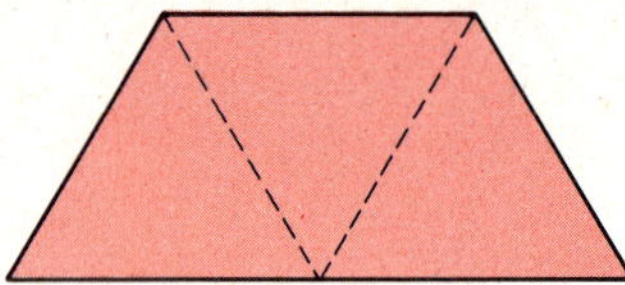

2 Cut out each net as a single piece.

3 Lightly score along the dotted lines.

4 Fold and glue the flaps to the outside to make TWO shapes like this

5 Now try to fit the TWO shapes together to make a triangular pyramid.

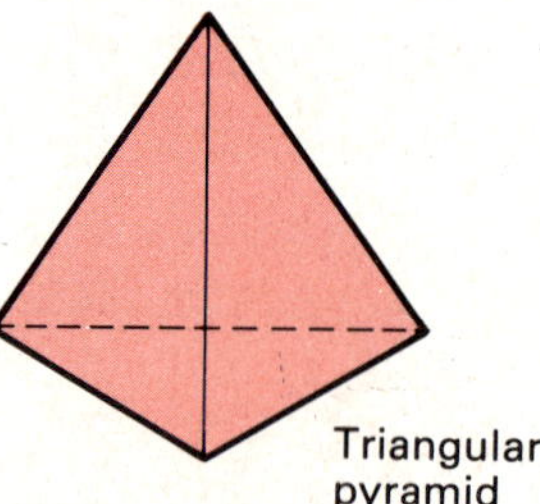

6 Ask your friends to try this puzzle.